Unintended Dystopia

Unintended Dystopia

RUSS WHITE

CASCADE *Books* · Eugene, Oregon

Cascade Books
An Imprint of Wipf and Stock Publishers
199 W. 8th Ave., Suite 3
Eugene, OR 97401

www.wipfandstock.com

PAPERBACK ISBN: 978-1-7252-7047-3
HARDCOVER ISBN: 978-1-7252-7048-0
EBOOK ISBN: 978-1-7252-7049-7

Cataloguing-in-Publication data:

Names: White, Russ, author.

Title: Unintended dystopia / Russ White.

Description: Eugene, OR: Cascade Books, 2021 | Includes bibliographical references.

Identifiers: ISBN 978-1-7252-7047-3 (paperback) | ISBN 978-1-7252-7048-0 (hardcover) | ISBN 978-1-7252-7049-7 (ebook)

Subjects: LCSH: Computer networks—psychological aspects. | Technology—Religious aspects—Christianity. | Theological anthropology—Christianity.

Classification: BT702 .W52 2021 (paperback) | BT702 (ebook)

12/06/21

Contents

Illustrations

Preface

I AM A GEEK. There is no getting around it, so I might as well start there. When people meet me, they assume I must have a plethora of social media accounts. After all, I work in and around the technologies that make social media networks possible. Further, I am an author, and authors need a platform. In the modern world, platforms mean social media.

So when people ask me about how to connect with me on social media, and I tell them that I only use social media for professional purposes, I do not have any social media apps on any device, and I only have accounts in a few select places, I normally receive some sort of blank stare. The answer seems to create a momentary pause. The following question is inevitable—why not?

This is where the journey to writing this book began. Why not? To give an answer, however, is to move far beyond the world of technology. Understanding why not requires beginning with the person and what being a person means. Technology is always both a tool and a shaper of a worldview. What does the tool of modern technology say about how we understand one another?

The answer is in the language we use, the way we find one another for relationships, the way we are presented and consume information, and the way we talk to one another. The answer is—we far too often see one another as objects or audiences, rather than as real people. There are plenty of books and research papers that show this is happening. The "what" is not really in dispute at this point. Countering the way we treat one another, however, requires much more than knowing the answer to "what." What is needed to understand "why."

This book is my attempt at explaining why. What is unique is that I am not a philosopher, although I have training in philosophy. I am not a theologian, although I have training in theology. I am not a cultural critic—well, maybe everyone is, but I am not trained in this area. I am not an ethicist.

What I am is a geek—a thirty-year veteran of the information technology industry. I helped to build some of the protocols and systems that run the Internet. I have built some of the most complex and largest computer networks in the world.

Therefore this book is important for you to read and understand. You may not completely understand what is happening when you are done, but you will have a much better sense of why it is happening.

Acknowledgments

It is difficult to know, across a lifetime of writing, exactly where this or that piece of information was picked up—especially with a work of this kind, which crosses so many fields of study. Going back twenty years, Brett Bartow at Pearson took a chance, launching my career in technical publishing that continues to this day. Chris Cleveland taught me style and grammar in his editing. Ethan Banks encouraged me to start writing again when I had stopped. Thousands of people have supported me in my work across these many years by buying things I've written and telling me how much my work means to them.

Three key areas of knowledge are represented in this book. The first is technical, an area I know well because of my thirty years building networks. I have learned from too many people to name them all, but a short list might include Don Slice, Alvaro Retana, and Ivan Pepelnjak. Theology I learned from Doug Bookman, Larry Pettegrew, and Robert Dean. Philosophy I learned under Bruce Little, Greg Welty, James Dew, and many others besides. Bruce Little and Doug Bookman, in particular, believed in me at the very beginning of my journey into a theological education, changing the very course of my life.

Combining these three fields into a whole that tries to make sense of neurodigital media, and how it is impacting our world, is a labor of years in research and writing. Huge measures of gratitude go to those who have supported me in many different ways, such as Hannah and Bekah ("shhhh, Dad's writing!"), Will Coberly, Tim Sigler, and the Pavilion of Men (you know who you are). My life is not without tragedy—the existential problem of evil is ever-present and very real to me—but the joy of a life lived among people who really love me is greater by any measure. The joy of God's love and truth is even greater still.

1

Utopia? Virtually So

In 1516, Sir Thomas More published a small book with the subtitle *Libellus vere aureus, nec minus salutaris quam festivus, de optimo rei publicae statu deque nova insula Utopia,* or "a little, true book, not less beneficial than enjoyable, about how things should be in the new island Utopia." The title of More's book is *Utopia*—a word that appropriately means "nowhere," because he believed no such place could exist. The global digital commons, made up of platforms such as Facebook, LinkedIn, Instagram, Google, Amazon, and others are sometimes considered a sort of utopia, as well. Many see these services as a world much like More's island, made new and fantastic—a virtual place built using information technologies early observers said would "empower the individual, enhance personal freedom, and radically reduce the power of the nation-state," where "existing social, political and legal power structures will wither away to be replaced by unfettered interactions between autonomous individuals and their software."[1]

Utopia, however, applies to this virtual world, the global digital commons, in another way. The digital world is both everywhere and nowhere at the same time. It is a place so constructed that everyone is seemingly just a few feet from everyone else on the Earth, and yet no one every truly touches one another. The very words used to describe modern information technology, *the cloud* and *serverless computing,* imply slipping from physical bounds into a place that is both nowhere and everywhere at the same time. This digital world is illusion—*the cloud is just other people's computers.*

1. Barbrook and Cameron, "The Californian Ideology."

1

This book is concerned with the three things used to create the illusion of a virtual world: digital computing, ways of organizing data (specifically the *social graph*), and User Experience (UX). Putting these three things together creates something called (in this book) *neurodigital media*. Neurodigital media is *neuro* because it interacts with user's beliefs and actions. Sometimes neurodigital media flattens the world, while other times it thickens the world—but it always makes the world appear differently than it otherwise might. Neurodigital media is *digital* because it relies on digital computing, and the way digital computers organize, analyze, and present information. Neurodigital media is *media* because it presents information and experiences (increasingly more experiences than information) to its users. This book focuses on three kinds of systems, or platforms, built using neurodigital media:

- Social media, such as Facebook, LinkedIn, and Instagram. These services allow users to make "connections" or "friends"; interact with other users through posts, games, and groups; allow users to "private message" one another; and curate a "content feed" for each user.

- Social recommenders, such as those used by Amazon and Netflix to suggest products or services a user might be interested in.

- Search engines, which act as a "gateway to the Internet" by allowing users to search for products or sources of information.

These are the kinds of services most users will encounter in their everyday lives, and they are also the kinds of services researched in order to gauge their impact. They will all be considered different forms of *neurodigital media* throughout this book, and the term *neurodigital media* will be used to denote and describe all three.

It is okay if you do not know what these terms mean. The first few chapters of this book will help you understand what they mean even if you struggle to use the electronic devices seemingly springing up like weeds in your life, within the context of their history. When you are finished reading this book, you will be able to understand these technologies, and their underlying presuppositions, well enough to ask a question of great importance: *what is gained, and what is lost, in using neurodigital technologies?* This critique will set the stage for a second volume that considers how Christians can respond to and responsibly use neurodigital media.

The Necessity of Critique

Why should there need to be a critique of these technologies? The hammer appears to be neutral; you can drive nails or break windows. The tool doesn't specify its use. But—what if there were no hammers? Could you build houses or bridges, raise a barn or lower a picture on the wall? Without houses, bridges, barns, and picture hanging, would our culture be the same? It does not seem so—as Craig Gay notes, "technological making is indissolubly linked with the distinctively human way of being-in-the-world."[2]

Are information technologies, such as neurodigital media, any different? Geert Lovink describes the relationship between information technology—which can be considered a superset of neurodigital media—and culture as institutional dogmas "hidden inside media folklore, hardwired into network architectures, steered by algorithms."[3] The motivation of a moment becomes fixed in culture over time, shaping the way future generations think and *are allowed to think*.

What is the purpose of these technologies? James Williams says they are designed to be "guiding lights, to help us navigate our lives in ways *we* want them to go."[4] Williams continues: "In a sense, our information technologies ought to be GPSes for our lives. (Sure, there are times when we don't know exactly where we want to go in life. But in those cases, technology's job is to help us figure out what our destination is, and to do so in the way we want to figure it out)."[5]

Where do we want to go? Where do these technologies lead? Why are some goods chosen to be "made possible," while others are chosen to become difficult or impossible? The following sections explore the promise and peril of technology.

Technophilia

No, do not bother looking in your dictionary; you will not find the word *technophilia*. The meaning of the word is simple to work out, however: *techno* relates to technology, and *philia* relates to love. Literally, then, technophilia is the love of technology—a condition that seems to pervade modern culture from the largest scale, such as government, to the smallest, the individual *technophile*. Why is modern culture so taken with technology? A

2. Gay, *Modern Technology and the Human Future*, 25.

3. Lovink, *Sad by Design*, loc. 458.

4. Williams, *Stand Out of Our Light*, 3.

5. Williams, *Stand Out of Our Light*, 9.

good place to begin in understanding the love of technology is in the first few pages of the Scriptures, Genesis 3:21.

In order to fully understand Genesis 3:21, however, the stage must be set by reviewing events up to this point. Genesis 3:1–6 tells of a series of events where Eve succumbs to the temptation of the snake, eating of the tree of the knowledge of good and evil. Adam, apparently valuing his relationship with Eve over his relationship with God, eats of the fruit as well (Genesis 3:6). The consequences of this single action, the Fall, are wide-ranging.

The woman is separated from the snake (Genesis 3:15); the woman is separated from her children, symbolized by pain in childbirth (Genesis 3:16); the woman is separated from her husband (Genesis 3:16); and the man is separated from the Earth (Genesis 3:17–18). Finally, the man and woman are both separated from God by being cast out of the garden, into the wilderness beyond, their way back in blocked (Genesis 3:24).

One further result of their action in eating of the fruit is that they now know they are naked (Genesis 3:7). God recognizes this problem, and provides clothes made from the skins of animals to cover their nakedness in Genesis 3:21: "And the Lord God made for Adam and for his wife garments of skins and clothed them." This is arguably the second instance of technology in the Scriptures. The first appears in Genesis 3:7b: "And they sewed fig leaves together and made themselves loincloths."

Two interesting points about this first use of technology in the Scriptures are important in laying a foundation for understanding the use of information technology. First, God does not disapprove of technology. Instead of God disapproving of technology, there are many instances of God instructing people in the use of technology throughout the Scriptures. Second, the use of technology in Genesis 3 is directly aimed at ameliorating the impact of sin.

Technology in the Scriptures

There are many examples of God teaching people how to use technology in the Scriptures, beginning with God teaching Adam and Eve how to make clothing out of animal skins in Genesis 3:21.

In Exodus 26–30, God gives Moses a detailed set of instructions about how to build the tabernacle God is going to dwell in while Israel is moving through the desert. While the instructions were (probably) sometimes new to the Israelites (such as the command not to adorn anything in the tabernacle with the form of any living thing, accompanied by detailed instructions on what to adorn the tabernacle with), some of them probably contained well-known concepts. For instance, God used *acacia wood* and *covered with gold* without any further instruction; these were apparently concepts the Israelites already knew.

One point about these passages stands out; in Exodus 31:1–6, God tells Moses:

> See, I have called by name Bezalel the son of Uri, son of Hur, of the tribe of Judah, and I have filled him with the Spirit of God, with ability and intelligence, with knowledge and all craftsmanship, to devise artistic designs, to work in gold, silver, and bronze, in cutting stones for setting, and in carving wood, to work in every craft. And behold, I have appointed with him Oholiab, the son of Ahisamach, of the tribe of Dan.

Not only has God told Moses what techniques to use—specifically how to embody a tabernacle in technology—he has even chosen and blessed specific members of the Israelite community to embody his instructions in a physical structure.

In Genesis 6, God instructs Noah to build an ark that will take him through the coming flood. God not only tells Noah to build this ark, he goes into detail explaining what materials and methods (techniques) to use in its building. It is to be made of gopher wood and covered with pitch inside and out (Genesis 6:14). The ark is to be 300 cubits long, 50 cubits wide, and 30 cubits high (Genesis 6:15). It is to be built with rooms, a roof, and three decks (Genesis 6:14–15). In 1 Chronicles 28, David gives the plans for a temple of God to Solomon to build. These plans, according to David, were given to him by God: "All this he made clear to me in writing from the hand of the Lord, all the work to be done according to the plan" (1 Chronicles 28:19).

Another line of evidence showing God is not opposed to technology is given by Andy Crouch. Cities, in the Scriptures and history, are often tied to the development and use of technology. It is densely populated areas focused on urban lifestyles that often encourage and harbor the invention and use of technology; for instance, Silicon Valley in California is well known for its emphasis on technical innovation.

On the other hand, Crouch notes that in reading Genesis, cities are always in the foreground of all that is evil: in 4:17, the first city is founded by a murderer; in chapter 11, the city of Babel rebels against God and must be destroyed; and in chapter 13, Abraham lives on the plains, while Lot moves to the city of Sodom, which is eventually destroyed for its sexual sins. By constantly relating cities to evil, the writer of Genesis also seems to cast aspersions on technology because cities are often where technologies are invented, and a focus of their use.

It is surprising, then, to read Revelation 21:2: "And I saw the holy city, new Jerusalem, coming down out of heaven from God, prepared as a bride adorned for her husband." According to Crouch, readers should be surprised by the existence of a *holy city*, given the record of Genesis in relation to cities.[6]

If God ultimately creates a *holy city* that is to become the dwelling place of many millions of saints, it is clear God is not opposed to technology.

Technology as a Counter to the Result of Sin

It is striking that many of the instances of technology in the Scriptures are aimed at a singular purpose: diminishing the impact of sin. In Genesis 3:21, God clothes Adam and Eve with clothes made from skin, which presumably means some animal was slaughtered in order to create these clothes; many scholars consider this the first sacrifice.[7] This first use of technology, then, not only covered the nakedness Adam and Eve experienced as a result of this first sin (Genesis 2:7), but also involved the first sacrifice to cover their sin. God, in providing a covering for Adam and Eve, offers a blessing rather than a curse.

The narrative of the flood in Genesis 6 is another example of God directly intervening in history by showing a human how to use technology. In Genesis 6:7–8, the writer describes the situation, saying: "So the Lord said, 'I will blot out man whom I have created from the face of the land, man and animals and creeping things and birds of the heavens, for I am sorry that I have made them.' But Noah found favor in the eyes of the Lord."

The coming flood, according to God, is a direct result of willful and continuous sin on the part of humanity. Because Noah finds favor, God instructs him to build an ark, providing the details of the ark's construction to ensure Noah has the technical know-how to survive the coming flood. In 1 Peter 3:19–21, Noah's survival through the flood is directly related to baptism:

> [Christ] went and proclaimed to the spirits in prison, because they formerly did not obey, when God's patience waited in the days of Noah, while the ark was being prepared, in which a few, that is, eight persons, were brought safely through water.

6. Crouch, *Culture Making*, 121.

7. See, for instance, Mathews, *Genesis 1–11:26*, 254.

> Baptism, which corresponds to this, now saves you, not as a re-
> moval of dirt from the body but as an appeal to God for a good
> conscience, through the resurrection of Jesus Christ.

In the narrative of the flood God once again provides a human with the technology required to diminish the result of sin.

A third, and final, example may be useful to show the point. There are two great construction projects in the Tanakh related to the worship of Yahweh. The first of these is the tabernacle, which was built while Israel was in the desert, after escaping from Egypt, and before finally entering the promised land. The description the Lord gives of the tabernacle is extensive, beginning in Exodus 25 and continuing through 31. Details such as the weight and size of each item, what each item was to be made of, how each item was to be put together and carried, are provided in these six chapters. In Exodus 31:1–11, God tells Moses he has appointed two people, Bezalel and Oholiab, to supervise and complete the work. This passage certainly seems to be an instance of God directly teaching specific techniques to build a building worthy of his presence.

Interestingly, a parallel set of events seem to occur in 1 Chronicles 28:11–19. The passage is shorter, but suggestive of the same level of care and detail on the part of God:

> Then David gave Solomon his son the plan of the vestibule of
> the temple, and of its houses, its treasuries, its upper rooms, and
> its inner chambers, and of the room for the mercy seat . . . "All
> this he made clear to me in writing from the hand of the Lord,
> all the work to be done according to the plan."

What is the specific purpose of these two buildings? They are to serve as a place where God's presence can reside with Israel. Not since the plac-ing of the angels guarding the way to the tree of life in Genesis 3:24 had God physically resided with humans. Why did God not reside with people after that time? God drove them out of his presence *because of their sin*. In this way, then, the tabernacle and temple were both technological artifacts designed to lessen a specific aspect of the result of sin.

Beyond this, however, these buildings were designed to be a place where sacrifices could be offered in order to cover the sins of individuals and the nation of Israel. In this way, then, these two buildings represent a second way in which the impact of sin was diminished.

The Promise of Technology

The modern view of technology begins in this point: technique, and its embodiment in technology, should be used—in the words of Brent Waters—to "make possible human mastery over time and place, and correlatively over nature and human nature."[8] J. B. Bury, in his study of the idea of progress, says of Francis Bacon:

> The principle that the proper aim of knowledge is the amelioration of human life, to increase men's happiness and mitigate their sufferings—*commodis humanis inservire*—was the guiding star of Bacon in all his intellectual labour. He declared the advancement of "the happiness of mankind" to be the direct purpose of the works he had written or designed. He considered that all his predecessors had gone wrong because they did not apprehend that the *finis scientiarum,* the real and legitimate goal of the sciences, is "the endowment of human life with new inventions and riches"; and he made this the test for defining the comparative values of the various branches of knowledge.[9]

After Bacon, however, the idea of using technology to diminish the impact of sin was unmoored from sin and God. Judeo-Christian belief, which gave birth to the idea of using technology to make humanity's lot more tolerable, became something to be escaped. J. B. Bury says:

> The idea of the universe which prevailed throughout the Middle Ages, and the general orientation of men's thoughts were incompatible with some of the fundamental assumptions which are required by the idea of Progress. According to the Christian theory which was worked out by the Fathers, and especially by St. Augustine, the whole movement of history has the purpose of securing the happiness of a small portion of the human race in another world; it does not postulate a further development of human history on earth.[10]

Once unmoored from God and sin, technology could then be moored again to the idea of progress; it is through advancing technology that progress could be made, ultimately perfecting even humanity itself:

> This doctrine of the possibility of indefinitely moulding the characters of men by laws and institutions—whether combined

8. Waters, *Christian Moral Theology*, 18.

9. Bury, *The Idea of Progress*, 51.

10. Bury, *The Idea of Progress*, 19.

or not with a belief in the natural equality of men's faculties—
laid a foundation on which the theory of the perfectibility of
humanity could be raised. It marked, therefore, an important
stage in the development of the doctrine of Progress.[11]

The technophile, then, looks not only to increasing the comfort and
survival of individuals against nature (which is no longer *fallen* but *harsh)*,
but the evolution of the entire human race.

The Promise of Neurodigital Media

What of neurodigital media? As a form of technology, neurodigital media
hold some promise to gain the kinds of widespread use resulting in a global
digital commons. Each manifestation of neurodigital media promises
slightly different benefits.

Social media promises increased social reach and economic benefits
to its users. LinkedIn's vision is to create "economic opportunity for every
member of the global workforce" by connecting "the world's professionals
to make them more productive and successful."[12] LinkedIn seems to fulfill
this role by being a source of leads and customers for businesses, as well
as a source of professional talent for recruiters.[13] These services are also a
major source of information for their users; in 2017, Elisa Shearer and Jeffrey Gottfried reported a majority of Americans use Facebook and Twitter
as sources of news.[14]

Users are not the only ones who see value in services like Facebook; the
company has grown from being worth a little over $200 billion in 2015 to
more than $600 billion in 2020.[15] What can a company sell to be worth this
amount of money? First, social media services can sell access to users—not
just *all* their users, of course, but rather carefully selected sets of users most
likely to purchase a product or change their opinion on a given topic (the
selection process is considered in more detail in a later chapter). This kind
of access allows advertisers to *target* their messaging (which some consumers prefer),[16] which means they can create more effective advertising campaigns. Second, social media services can sell (theoretically) anonymized

11. Bury, *The Idea of Progress*, 167.

12. "About LinkedIn."

13. Osman, "Mind-Blowing LinkedIn Statistics and Facts (2019)."

14. Shearer and Gottfried, "News Use Across Social Media Platforms 2017."

15. Zacks Investment Research, "Facebook, Inc. (FB) Market Cap."

16. Knight, "6 Reasons."

data about their users. Elena Botella says Facebook earns around $130 each year from the data it has on each user.[17] User tracking and advertising are often justified as a good tradeoff for free content available on the global digital commons.[18]

Social recommenders allow organizations to directly target users with content they will find interesting or useful in some way. Netflix and You-Tube use social recommenders to suggest the next movie or video to watch; Amazon and other vendors use these systems to suggest additional items to add to your shopping cart. These systems can help users discover products they might enjoy or prevent them from forgetting something they might want or need (such as always buying conditioner with shampoo, or cream cheese with bagels), so they often do serve the user in useful ways. Recommender systems also increase revenue streams and certainty for the companies that use them. According to *Rejoiner,* Amazon increased its retail sales by 29 percent in 2015[19] by identifying consumer preferences through past purchases and the social environment. Shabana Arora states Netflix successfully captures the attention of consumers to drive subscriptions by personalizing recommendations.[20]

Search engines, the third kind of neurodigital media-based system considered here, power the entire Internet, in a sense. The ability to find information across the entire global digital commons from a single point adds immense value to the lives of millions of users billions of times a day.

The ability to understand and shape human behavior through neurodigital media has applications far beyond personal convenience and commercial applications. In November of 2015, a series of coordinated terrorist attacks rocked Paris. Three suicide bombers set off explosions outside the Stade de France, followed by several mass shootings at public places, including a large concert. One-hundred and thirty people died; more than four hundred others were injured. Rob Enderle, writing after these events, says ad profiles could (and should) be used to profile people attending any event and "used to risk assess everyone profiled and suggest approaching behaviors, which could result in unfortunate violence."[21] After outlining several other ways these kinds of technologies could be used to promote public safety, he turns to more mundane possibilities—"you'd also know the kind of in-depth information on an employee that would make them

17. Botella, "Facebook Earns $132.80 From Your Data per Year."
18. AdExchanger, "If A Consumer Asked,"
19. Rejoiner, "Amazon's Recommendation Engine."
20. Arora, "Recommendation Engines."
21. Enderle, "In the Shadow of Paris."

happier and perform better."[22] Cao Zhihui touts the crime-fighting capability of systems that can assess, predict, and shape human behavior: "It might not yet be possible to build a city with no shadow areas, but advancements in AI and increasing digitization mean it won't be long before we can use this technology to deduce criminals' intentions and the tricks of the trade with the crime-fighting AI."[23]

The applications of neurodigital media in its various forms to improve the human condition are seemingly endless. Modern democratic states often emphasize the power of people to select leaders and policies, but countries with hundreds of millions of citizens and thousands of complex policies pose a challenge to this vision. According to Joe Trippi, social media can be applied directly to these problems, restoring the "primacy of the individual voter."[24] As noted earlier, James Williams summarizes the perceived benefits of using systems that can predict preferences and guide behavior to solve social problems, saying they are designed to be "guiding lights, to help us navigate our lives in ways *we* want them to go."[25] Williams continues: "In a sense, our information technologies ought to be GPSes for our lives. (Sure, there are times when we don't know exactly where we want to go in life. But in those cases, technology's job is to help us figure out what our destination is, and to do so in the way we want to figure it out)."[26]

Technophobia

While many futurists look forward to the day when every piece of clothing, every appliance, every door, every vehicle, and even the very trees have sensors, there is a long history of others worrying about what all these technologies are doing to people and societies. Such criticisms of new technologies reach all the way back to Plato, who in *Phaedrus* comments on the invention of writing, saying:

> [Y]ou who are the father of letters, from a paternal love of your own children have been led to attribute to them a quality which they cannot have; for this discovery of yours will create forgetfulness in the learners' souls, because they will not use their memories; they will trust to the external written characters and not remember of themselves. The specific which you

22. Enderle, "In the Shadow of Paris.".
23. Zhihui, "Nowhere to Hide," 54.
24. Trippi, "Technology Has Given Politics Back Its Soul," 34.
25. Williams, *Stand Out of Our Light*, 3.
26. Williams, *Stand Out of Our Light*, 9.

have discovered is an aid not to memory, but to reminiscence, and you give your disciples not truth, but only the semblance of truth; they will be hearers of many things and will have learned nothing; they will appear to be omniscient and will generally know nothing; they will be tiresome company, having the show of wisdom without the reality.[27]

More recently, fears of awful uses of technology have found their way into the popular imagination through fiction. For instance, in Ray Bradbury's *Fahrenheit 451*, Guy Montag's wife, Mildred, is obsessed with watching television, burying herself in a world of commercials and entertainment rather than conversing with her husband. Bradbury describes a "television room" in the house, probably modeled on early restaurants (such as the Varsity in downtown Atlanta, Georgia) where a person can be surrounded by television sets.[28] Mildred calls the people on this immersion experience her "family," refusing to talk to her husband. While the television draws people into a dream, the mechanical hound drives them from books and reality.[29]

In Aldous Huxley's *Brave New World*, another widely read dystopian novel, technology has reached the point of producing humans with characteristics fit for individual castes. This produces a society where everyone is perfectly fit for their jobs and stations in life, production and consumption are maximized, and pharmaceuticals are always available in those cases where someone is feeling out of sorts. In George Orwell's *1984*, technology plays a role closer to the mass surveillance many fear in the modern world: "Big Brother is always watching." In Yevgeny Zamyatin's *We*, written in 1924, a spacecraft engineer lives in a world where people are not given names, and the buildings are (mostly) made of glass in order to enable full surveillance of every person (similar to the concept of a panopticon, as described by Jeremy Bentham.)[30]

Moby and The Void Pacific Choir, in their official video of the song "Are You Lost In The World Like Me?," offer an even bleaker image of the world of technology. A young boy wanders through the world trying to get anyone's attention—pulling on people's shirts and waving his arms in their faces. Everyone in the video stares into their phones, paying no attention to where they are going. In one scene, people walk into an uncovered manhole; in the final scenes, they walk over a cliff. In both cases, they do not seem to care because they are entranced by the faint blue glow of a tiny screen.

27. Plato, *The Dialogues of Plato*, 274.
28. Bradbury, *Fahrenheit 451*, 44.
29. Bradbury, *Fahrenheit 451*, 22.
30. Bentham, *The Works of Jeremy Bentham*, 4:37.

In one of the final scenes, a girl is filmed dancing wildly, and then bullied, finally committing suicide by jumping off a tall building. The people in the street film her with their phones, not seeming to care that the girl is falling to her death.[31]

There are plenty of arguments against technology outside the imaginative realm, as well. Nicholas Carr argues that the widespread use of technology—specifically what is called neurodigital media here—changes the physical "wiring" of the user's mind, much like books did in centuries past. While the changes the book made are positive, the changes wrought by modern information technologies are not, according to Carr:

> Our brains become adept at forgetting, inept at remembering. Our growing dependence on the Web's information stores may in fact be the product of a self-perpetuating, self-amplifying loop. As our use of the Web makes it harder for us to lock information into our biological memory, we're forced to rely more and more on the Net's capacious and easily searchable artificial memory, even if it makes us shallower thinkers.[32]

This impact on the human brain is paralleled by changes in the way people consume and understand information. Peter Pomerantsev argues that online spaces in the form of social media induce distortion: "It really doesn't matter if stories come from dodgy sources: you're not looking to win an argument in a public space to a neutral audience; you just want to get the most attention possible from like-minded people. Indeed the more extreme position you take, the better."[33]

Information technology not only impacts the way people think, it also impacts what people think about. Nolen Gertz argues that one of the most widely used social media platforms, Facebook, has four components: misinformation, manipulation, dependency, and distraction.[34] He likens the hyperscale social media site to a pair of glasses; while it helps people focus on particular bits of information better, and brings things into focus that could not previously be seen, it also distorts others, making them impossible to see. Robert Epstein and Ronald Robertson show that the ranking of pages within search results has a significant impact on the beliefs people using the search engine form through the primacy effect.[35] Just as the first candidate listed on a ballot will likely garner more votes, the first "hit" in the results

31. Moby and The Void Pacific Choir, "Are You Lost In The World Like Me?"
32. Carr, *The Shallows*, loc. 3321.
33. Pomerantsev, "Disinformation All the Way Down."
34. Gertz, "The Four Facebooks."
35. Epstein and Robertson, "The Search Engine Manipulation Effect."

of a search on the Internet will likely garner more "clicks." This power of search engines has been linked to influencing everything from purchasing decisions to the outcomes of elections.

Search engines rely on grouping searches—if many people look for two products in quick succession, the search engine will "learn" this pattern, giving results for the second product when someone searches for the first. This kind of patterning extends far into all social recommender systems, including applications that recommend products, music, directions, places to eat, and things to do. This is called the filter bubble, and it has been shown to produce large changes in the way people see and understand the world.[36]

Neurodigital media also appears to be harmful to relationships. Valentina Rotondi and others argue that "due to its intrusiveness, the smartphone reduces the quality of face-to-face interactions and, as a consequence, their positive impact on well-being."[37] Social media can have direct impact on relationships as people perceive the decisions of their "in-group" as validation of their own decisions. Nellie Bowles argues that the social fabric of modern technology-based societies are altering to the point that human contact is becoming a luxury good. After describing an older person who has adapted his life to an on-screen animated character, she says:

> The rich do not live like this. The rich have grown afraid of screens. They want their children to play with blocks, and tech-free private schools are booming. Humans are more expensive, and rich people are willing and able to pay for them. Conspicuous human interaction—living without a phone for a day, quitting social networks and not answering email—has become a status symbol. All of this has led to a curious new reality: Human contact is becoming a luxury good.[38]

The culture created by social media, Lovnik declares, creates "experience junkies who desire to wring out life's pleasures, to thoroughly exhaust it," who "want so much, and make so little."[39] Wendy Hui Kyong Chun describes the results of social media as "forever trying to catch up, updating to remain (close to) the same; bored, overwhelmed, and anxious all at once . . . What matters most: figuring out what will spread and who will spread it fastest."[40]

36. DuckDuckGo, "Measuring the Filter Bubble."

37. Rotondi, Stanca, and Tomasuolo, "Connecting Alone."

38. Bowles, "Human Contact Is Now a Luxury Good," March 2019.

39. Lovink, *Sad by Design*, loc. 285.

40. Chun, *Updating to Remain the Same*, loc. 176.

The Question of Technological Determinism

Accepting either technophilia or technophobia requires accepting that technology can cause people and cultures to shape themselves around technology. Some hold the widespread use of a technology can shape culture, allowing either the promises of technophilia or the perils of technophobia to be realized. Are technology and culture tied together tightly enough to support the connection between neurodigital media, individual lives, and the larger culture contemplated here?

Technological Determinism

According to Merritt Roe Smith and Leo Marx, technological determinism as a term may not have explicit meaning for the average person in a modernized society, but the "steady growth of [technological] power is just another self-evident feature of modern life, an obvious fact that calls for no more comment than the human penchant for breathing."[41] Smith and Marx argue the defining characteristic of technological determinism is the perception of "a vivid sense of the efficacy of technology as a driving force of history: a technical innovation suddenly appears and causes important things to happen."[42] Smith separately argues the belief that "technology as a key governing force in society dates back at least to the early stages of the Industrial Revolution."[43]

Examples of thinkers who held to technological determinism early in the Industrial Revolution include Thorstein Veblen, who argued in 1906 that "machine technology has made great advances . . . and has become a cultural force of wide-reaching consequence," and: "The machine process has displaced the workman as the archetype in whose image causation is conceived by the scientific investigators."[44] Charles Beard, writing in 1930, says technology makes the social environment "highly complex," replacing a "simple order of farmers and merchants" with a "highly specialized society composed of engineers, machinists, bacteriologists," and others.[45] Marx ties the technology used in the production of goods (what he calls the mode of production) to the form of government, and from there implicitly to the structure of society, saying:

41. Smith and Marx, "Introduction," ix.
42. Smith and Marx, "Introduction," x.
43. Smith, "Technological Determinism in American Culture," 2.
44. Veblen, "The Place of Science in Modern Civilization," 597.
45. Beard, *The American Leviathan*, 6.

Social relations are closely bound up with productive forces.
In acquiring new productive forces people change their mode
of production; and in changing their mode of production, in
changing the way of earning their living, they change all their
social relations. The hand mill gives you society with the feudal
lord; the steam mill, society with the industrial capitalist.[46]

In 1944, during the course of World War II, Charles Dawson took a
less elevated view, tying technology to the rise of mass society, where the
state takes the "peasant from his bullock-cart and teaches him to drive an
automobile," ultimately leading to his integration into a "mass that is being
irresistibly impelled forward in a race of industrial production or military
destruction."[47] Dawson continues, saying this mass society is "hostile to
freedom, since it reduces the control of the individual over his own life and
makes him the instrument of collective forces on which his very existence
depends."[48]

In 1964, examining the impact of the previous decades of the Industrial Revolution on culture, and looking ahead into an even more technologically focused future, Jacques Ellul argued:

But when technique enters into every area of life, including the
human, it ceases to be external to man and becomes his very
substance. It is no longer face to face with man but is integrated
with him, and it progressively absorbs him. In this respect, technique is radically different from the machine. This transformation, so obvious in modem society, is the result of the fact that
technique has become autonomous.[49]

Technique becoming autonomous means technology itself acts as a
force which drives culture and cultural change without human interven-
tion. Ellul emphasizes the autonomous power of technology when he says
"the machine," if given its head, will topple "everything that cannot support
its enormous weight."[50] In the same year, Marshall McLuhan linked media
technologies to culture, saying "the medium is the message," and the "re-
structuring of human work and association was shaped by the technique of
fragmentation that is the essence of machine technology."[51] The argument

46. Marx, *The Poverty of Philosophy*, 122.
47. Dawson, "Religion and Mass Civilization," 2.
48. Dawson, "Religion and Mass Civilization," 2.
49. Ellul, *The Technological Society*, 6.
50. Ellul, *The Technological Society*, 5.
51. McLuhan, *The Essential McLuhan*, 151.

that technology drives cultural change cannot stand on its own, however; some mechanism must exist whereby technology can have an impact on culture. Robert Heilbroner says that for technology to drive culture, a "huge variety of stimuli, arising from alterations in the material background, must be translated into a few well-defined behavioral vectors."[52] Many different causal connections tying culture to technology have been suggested; three are considered here: economics, social organization, and the impact of technology on the thinking processes and skills of individuals within the culture.

Economics is the first of the three suggested causal connectors between technology and culture. Economics, according to Robert Heilbroner, is a "force field" that optimizes the creation of new wealth, and works by "ignoring all effects of the changed environment except those that affect our maximizing possibilities."[53] Changes in technology that maximize the possibilities of financial gains are, therefore, emphasized through "changes in the price system, indicating directions in which economic activity can most advantageously move."[54] Andy Crouch gives an example of this connection through the cultural importance of rivers versus highways, and the cities which are connected to each. According to Crouch, before the widespread use of automobiles, the courses and locations of major rivers were widely known, and cities were situated alongside rivers to facilitate transportation. Once use of the automobile became widespread, however, the culture shifted to emphasizing the course and location of major highways, and cities are situated in sites advantageous to highway access and automobile travel.[55] Technology, then, can change culture by changing the way the world is seen, as well as what is perceived to be important or unimportant.

A second way in which technology can drive culture is by changing the goals and organizing principles of government and civil organizations. Since the 1960s, the rise of information technology has led to unprecedented forms of social organization,[56] what Shoshana Zuboff calls an information civilization.[57] Thus, the modern world represents an entire culture founded on technology, in which J. B. Bury argues the "warrior, priest, and political leader sink into the background," while events operate "only in accordance

---

52. Heilbroner, "Technological Determinism Revisited," 71.
53. Heilbroner, "Technological Determinism Revisited," 72.
54. Heilbroner, "Technological Determinism Revisited," 72.
55. Crouch, *Culture Making*, 26.
56. Zuboff, *The Age of Surveillance Capitalism*, 12.
57. Zuboff, *The Age of Surveillance Capitalism*, 4.

with the economic realities produced by the machine."[58] Technology, according to Bury, "reinforces the social, as distinguished from the individual, aspects of historical evolution."[59]

Finally, technology can directly impact the way users of technology think. For instance, Thorstein Veblen, writing in 1904, says the "machine throws out anthropomorphic habits of thought," forcing the worker to adapt to the work, rather than the work to the worker.[60] Marshall McLuhan argues, in a similar vein, that: "Students of computer programming have had to learn how to approach all knowledge structurally," shaping the way they information so the computer can store and process it.[61] Nicholas Carr's book *The Shallows* is an extended argument on how using technology, particularly the Internet, changes the way users think at a fundamental, physical level through *neuroplasticity*. Carr says: "We become, neurologically, what we think."[62] Carr later ties these neurological changes to changes in culture wrought by technology: "Neuroplasticity provides the missing link to our understanding of how informational media and other intellectual technologies have exerted their influence over the development of civilization and helped to guide, at a biological level, the history of human consciousness."[63]

Against Technological Determinism

Technological determinism is well supported by the arguments above, but it is not impervious to criticism. According to Bruce Bimber, there are three kinds of technological determinism: normative, unintended consequences, and nomological.[64] Bimber argues that in the normative account of technological determinism, technology becomes "the norm of practice" because the technological "goals of efficiency or productivity become surrogates for value-based debate over methods, alternatives, means, and ends."[65] Against this view, Heilbroner notes that technological advances depend on the "rewards, inducements, and incentives offered by society," so "the direction of technological advance is partially the result of social policy."[66] Placing the

58. Bury, *The Idea of Progress*, loc. 270.

59. Bury, *The Idea of Progress*, loc. 270.

60. Veblen, *The Theory of Business Enterprise*, 310.

61. McLuhan, *The Essential McLuhan*, 90.

62. Carr, *The Shallows*, loc. 633.

63. Carr, *The Shallows*, loc. 873.

64. Bimber, "Three Faces of Technological Determinism," 81.

65. Bimber, "Three Faces of Technological Determinism," 82.

66. Heilbroner, "Technological Determinism Revisited," 62.

goals of efficiency or productivity above other possible goals is a cultural decision that drives the kinds of technologies developed and how those technologies are used. It is the value the culture places on efficiency of productivity, rather than any particular technology, which causes these values to be prized above all others.

The unintended consequences account of technological determinism, according to Bimber, is that "even willful, ethical social actors are unable to anticipate the effects of technological development. For this reason, technology is at least partially autonomous."[67] This view of technological determining, according to Bimber, fails because the adoption of technology has indeterminate results, rather than determinate.[68]

Bimber describes the nomological account as the laws of nature acting on the past and present in a way that only permits one possible future.[69] In these accounts, "the technology driven society emerges regardless of human desires and values."[70] This account of technological determinism, however, cannot seem to answer the challenge Lewis Mumford puts forward:

> All the critical instruments of modern technology—the clock, the printing press, the water-mill, the magnetic compass, the loom, the lathe, gunpowder, paper, to say nothing of mathematics and chemistry and mechanics—existed in other cultures. . . . They had machines; but they did not develop "the machine." It remained for the peoples of Western Europe to carry the physical sciences and the exact arts to a point no other culture had reached, and to adapt the whole mode of life to the pace and the capacities of the machine. How did this happen? How in fact could the machine take possession of European society until that society had, by an inner accommodation, surrendered to the machine?[71]

If technology drives changes in culture in a deterministic way, then the invention or widespread use of a technology within a culture should always cause the same kind of changes in that culture. The historical record, however, does not show this kind of correlation, so it would not seem technology is an autonomous driver of cultural change in the way technological determinism requires.

---

67. Bimber, "Three Faces of Technological Determinism," 85.
68. Bimber, "Three Faces of Technological Determinism," 89.
69. Bimber, "Three Faces of Technological Determinism," 83.
70. Bimber, "Three Faces of Technological Determinism," 84.
71. Mumford, *Technics and Civilization*, 4.

If Bimber's arguments against technological determinism stand, the application of technology is purely an artifact of the culture in which it is developed. In this view, technology may be said to be neutral. Langdon Winner says of this view: "Because technological objects and processes have a promiscuous utility, they are taken to be fundamentally neutral as regards their moral standing."[72] The danger of this view, according to Winner, is that it results in technological somnambulism, where the impact of technology on culture is ignored as we "sleepwalk through the process of reconstituting the conditions of human existence."[73] There are weaknesses in Bimber's argument, however, that require further consideration.

Bimber's criticisms treat each culture as a monolithic whole. Cultures, however, are composed of many subcultures, such as families, religious organizations, businesses, and governments with overlapping spheres of power. The impact of the adoption of any technology on a culture is going to depend on many factors, such as the rate at which technological changes can spread throughout a culture and how subcultures interact economically. Two examples from the Scriptures may be useful to illustrate this point. According to Jeremiah 35, the Rechabites were a clan that had taken an oath not to drink, build houses, or participate in agriculture of any kind. According to some scholars, the position of the Rechabites within Israel is parallel to that of marginalized metalworkers in other cultures.[74] The social stature of such a subculture would prevent any unique skills they possessed from being passed through to the larger culture regardless of their economic value, hence preventing them from enabling large-scale cultural changes. Genesis 46:34 provides another example of an occupation being considered marginal within a culture: "Your servants have been keepers of livestock from our youth even until now, both we and our fathers . . . [F]or every shepherd is an abomination to the Egyptians." The social status of shepherds in this culture would not be likely to encourage the spread of any skills unique to shepherding throughout the larger culture.

In instances, however, where subgroups within a culture compete economically, and one subculture deploys technology in a way that creates economic advantage, it is likely the larger culture will adopt these technologies. The widespread adoption of a technology, in this case, would have an impact on the larger culture, rather than being restricted in its application. Media technology was developed and applied during a time when technological advancement was widely seen as the reason for the survival of

72. Winner, *The Whale and the Reactor*, 5.

73. Winner, *The Whale and the Reactor*, 9–10.

74. McNutt, "The Kenites, the Midianites, and the Rechabites," 118.

Western cultures through a series of actual wars as well as the Cold War. According to Fred Turner, "the research laboratories of World War II and later, in the massive military engineering projects of the Cold War, scientists, soldiers, technicians, and administrators broke down the invisible walls of bureaucracy and collaborated as never before," embracing computers and imagining institutions as "living organisms."[75] The result was "self-evident financial and social success," widely seen as laying the foundation for a new economy.[76]

Bimber's arguments may rule out a strong form of technological determinism where technology is completely autonomous from culture, but his arguments do not address more nuanced views of the interaction between technology and culture.

Technological Momentum

Thomas P. Hughes describes an alternate way of understanding the relationship between technology and culture that he calls technological momentum. Hughes argues that technologies and the social systems surrounding these technologies form a technological system in which the technology and culture adapt to one another over time.[77] While technology forms the core of these systems,[78] the system becomes more social and less technical as it matures.[79] As such a system gains momentum, it becomes "less shaped by and more the shaper of its environment."[80] Andy Crouch argues that the influence of technology is in the creation of physical artifacts that then enable or restrict future generations in their range of available choices:

> We don't make Culture, we make omelets. We tell stories. We build hospitals. We pass laws. These specific products of cultivating and creating—borrowing a word from archaeology and anthropology, we can call them "artifacts," or borrowing from philosophy, we can call them "goods"—are what eventually, over time, become part of the framework of the world for future generations.[81]

75. Turner, *From Counterculture to Cyberculture*, loc. 110.
76. Turner, *From Counterculture to Cyberculture*, loc. 152.
77. Hughes, "Technological Momentum," 103.
78. Hughes, "Technological Momentum," 105.
79. Hughes, "Technological Momentum," 105–6.
80. Hughes, "Technological Momentum," 108.
81. Crouch, *Culture Making*, 28.

Television is a useful example of a technological system that has required the creation of a supporting subculture, has been shaped by both this subculture and the "external" culture, and has also shaped the "external" culture.

From today's perspective, the rise of television as a widespread medium dedicated to entertainment appears obvious and inevitable, but this appearance may primarily be due to survivor bias.[82] In the initial years of its broadening use, television was perceived as a potential educational technology, rather than entertainment.[83] It is possible television would have been restricted to educational uses if the price of televisions, combined with the increasing prosperity of families in the post-World War II era, had not resulted in widespread television ownership.[84] As history played out, however, the widespread ownership of television pushed broadcasters to find something more profitable than education, and the resonance of television drew programming towards entertainment. An entire social system of producers, performers, advertisers, measurement systems, engineers, suppliers, researchers, etc., was built around this new technology and the entertainment it delivered. The culture and resonance of television worked together to create a new artifact through which culture was then shaped.

If You Haven't Found the Trade-Offs . . .

Then you haven't looked hard enough. Far too often, the newly forming global digital commons are treated as an absolute good—but many writers and researchers believe there is danger in this new virtual world. The promise of "human mastery over time and place, and correlatively over nature and human nature,"[85] needs to be clearly set off against C. S. Lewis's warning that "what we call Man's power over Nature turns out to be a power exercised by some men over other men with Nature as its instrument."[86] The problem, ultimately, is *sin;* while technology may be able to ameliorate the effects of sin, it cannot eliminate sin itself. Regardless of how powerful individual human beings may become, or a culture within its historical setting may become, humans are still very much limited and beholden to their passions.

While many are familiar with the legend of the emperor with no clothes, not many are familiar with the legend of the clothes with no

82. Shermer, "How the Survivor Bias Distorts Reality."

83. Novak, "Predictions for Educational TV in the 1930s."

84. Clifton, "How Did World War II Affect Television?"

85. Waters, *Christian Moral Theology*, 18.

86. Lewis, *The Abolition of Man*, 55.

emperor. In a faraway land there lived a people who were quite proud of their tailors. The emperor of this land, being the chief citizen, was always the most finely dressed man in the land. One day, some tailors told him they could make him even finer before his people by creating a new suit every day for a full year—the emperor would never be out of style! The only problem was, of course, these suits were so fine the emperor would need to be sewn into each one, requiring many hours of work. Since the emperor was a busy man, he decided on a unique course of action: he would place each new suit, for the entire year, over the older suit from the day before. In this way he could save several hours each day, while also getting the new clothes he so deeply craved.

As the year passed, many wondered if the emperor were gaining too much weight, or if he were perhaps sick with some sort of disease. The royal doctors prescribed a diet, which the emperor—not a small man at the beginning of the year—stuck to fastidiously. At the end of the year, on the final day, as the tailors were putting the final touches on their very last suit, they bid the emperor to come out onto the balcony so everyone could see their finely wrought creation. The emperor did not move.

Worried, the tailors began stripping off layers of clothes, one at a time. Days later, they reached the bottom layer and discovered, much to their dismay, an empty shell. Their clothes had no emperor. Analogously, once the layers of digital identity are piled on over the coming years, the question worth asking is: will there be anyone left? Will the digital world have no people? If not, where will they have gone?

2

Intellectual Roots

FACEBOOK, GOOGLE, AND OTHER systems based on neurodigital media did not "spring from the minds of men wholly formed," like some Greek god. Instead, neurodigital media is a culture and technologies intertwined to form a whole. The culture of neurodigital media has four distinct components—the radical autonomy of the person, the engineering mind-set, the idea of social progress through technology, and a naturalistic view of the world, each of which has a history. These histories merged in two distinct groups in the 1950s, the Beats, and the research community focused on winning World War II. The Beats flowered into the hippies, and the research community flowered into the cold warriors, both centered in Southern California. The unlikely mixture of the hippies and the cold warriors resulted in the Californian Ideology, the ideology of neurodigital media. This chapter traces the intellectual roots of neurodigital media through time, beginning with the rise of radical autonomy.

Power to the Person

In 1562, the forces of the Duke François de Guise massacred a hundred Protestants attending a service of worship in a barn in the town of Wassy, France. While religious wars in Europe go back to the time of the Roman Empire's Constantinian synthesis, this singular event marks the beginning of 150 years of bloody wars between Catholics and Protestants, including the slaughter of the Huguenots and the Eighty Years' War. These wars were often ended by the state stepping in to create a protective hedge around groups, allowing them to hold to beliefs without interference and transferring trust from churches to governments. Over time, this protective hedge moved from the religious group to the individual, eventually morphing into the full-blown individual autonomy characterizing the Californian Ideology—and presupposed in modern implementations of neurodigital media.

Individual instances of the government stepping in to resolve religious disputes during this 150 years of bloodshed serve to illustrate the general trend. The Edict of Nantes, signed in 1598, although never fully followed (it was ultimately revoked in 1685 by the Edict of Fontainebleau), was one of the first attempts of a secular government to resolve the battles between the Protestants and Catholics. The Edict specified the French Huguenots would be afforded safe havens maintained by the French monarch, along with a series of places of refuge supported by the Huguenots. At the time of its signing, the general rule was *cuius regio, eius religio*—the people of a nation would follow the religious belief of the king.

In 1688, ninety years later, the nobles of England, Scotland, and Ireland revolted against King James in the almost bloodless Glorious Revolution. Leading up to the Revolution, King James stoked the fears of the Protestants by creating a court to settle religious disputes and disarming Protestants while arming Catholics, actions raising the specter of a new religious war. The immediate cause of the revolution was the birth of King James's first son, James Francis Edward, setting the stage for a Catholic dynasty. The parliament of England offered the crown to William of Orange and Mary II jointly, on the condition they would sign a Bill of Rights—one of the first such instruments in history. This Bill placed the power of the parliament over the crown in several essential respects, making it illegal for the monarch to create a court that could interfere in religious matters. The result was a sphere of autonomous freedom around each person; within the bounds of the Catholic and Anglican churches, individuals could believe and worship as they liked.

This trend towards individual autonomy continued in North America with religious groups seeking freedom from government interference in their belief and worship. The colonies fought a war to free themselves from the power of the King of England because they believed they should have equal representation in parliament ("no taxation without representation"). The cause of the colonies was tightly bound to individual consciousness in religious belief. Elisha Williams, writing in 1744, said:

> That the sacred scriptures are the alone rule of faith and practice
> to a Christian, all Protestants are agreed in; and must therefore
> inviolably maintain, that every Christian has a right of judging

for himself what he is to believe and practice in religion according to that rule: Which I think on a full examination you will
find perfectly inconsistent with any power in the civil magistrate
to make any penal laws in matters of religion. Tho' Protestants
are agreed in the profession of that principle, yet too many in
practice have departed from it.[1]

Williams continued, "Reason teaches us that all men are naturally
equal in respect of jurisdiction or dominion one over another," and "we are
born free as we are born rational."[2]

The individual's autonomy of belief primarily flows from the work
of John Locke, an English thinker who stood on the side of the noblemen
against King James in the Glorious Revolution. Locke argued that each person should seek out the truth through reason rather than accept the opinion
of authorities and apportion belief to the evidence available. Locke held that
citizens form governments to protect the life, liberty, health, and property of
individuals—an idea that made its way into the United States Constitution
in 1789.

In the United States, the seed of religious liberty grew into individual
autonomy through a right to privacy. In an opinion considering placing
limits on abortion, Supreme Court Justice Anthony Kennedy argued,
"At the heart of liberty is the right to define one's own concept of existence,
of meaning, of the universe, and of the mystery of human life."[3] The culture
of California, in particular, was heavily influenced by individualism beginning in the great Gold Rush and emphasized in the location of independent
movie studios in Los Angeles to avoid Thomas Edison's patents.[4] Autonomy
flowered in San Francisco with the 1960s' hippie movement, which emphasized communal living and sought "meaning in life that is personally
authentic."[5]

A New Set of Mind

In the ancient world, humans perceived individuals and entire cultures as
being at the mercy of nature. Gaining autonomy would require humankind
to have the ability to shape nature to its will. To obtain this power requires

1. Williams, *The Essential Rights and Liberties of Protestants*, 1.

2. Williams, *The Essential Rights and Liberties of Protestants*, 2.

3. Justia Law, *Planned Parenthood of Southeastern Pa. v. Casey.*

4. Aberdeen, "The Edison Movie Monopoly"; Lewis, "Thomas Edison Drove the
Film Industry to California."

5. "Port Huron Statement."

more than just a set of tools and techniques—it requires a new set of mind that will consistently seek out, find, and create new ways of controlling nature. The origins of this new mind-set are deeply bound up in the historical dispute between empiricists and rationalists.

In the twelfth century, William of Ockham planted the seed of a revived nominalism, which holds that words refer to reality, but they do not truly explain the world as it is. Words are not intrinsically connected to reality. Ockham's argument was widely accepted, resulting in a newly found gap between the mind and the world—a problem that vexes philosophy from Ockham's time to the present. Servais Pinckaers says:

> With Ockham we witness the first atomic explosion of the modern era. The atom he split was obviously not physical but psychic. It was the nadir of the human soul, with its faculties, which was broken apart by a new concept of freedom. This produced successive aftershocks, which destroyed the unity of theology and Western thought. With Ockham, freedom, by means of the claim to radical autonomy that defined it, was separated from all that was foreign to it: reason, sensibility, natural inclinations, and all external factors. Further separations followed: freedom was separated from nature, law, and grace; moral doctrine from mysticism; reason from faith; the individual from society.[6]

If such a gap exists between the mind and the real world, how can anyone know anything about reality? Rationalists, such as René Descartes, held the senses are imperfect, so the only way to know about the real world is through the mind. To find truth, you must begin with facts or ideas, rather than with experiences. One result of rationalism is that the mind and person are radically autonomous. While rationalism was eventually set aside because of the mind/body problem, radical autonomy remained, ultimately leading to nihilism, which is explored more fully in the next chapter.

Contrary to rationalists, empiricists like Francis Bacon held the senses might be unreliable, but the senses were the only path humans have to know nature. To resolve this difficulty, Bacon argued for using a specific method and specially made tools to measure with accuracy beyond what the senses could achieve. Bacon's argument led to the formulation of the scientific method, which promised not only to understand but control nature. The radical autonomy of the nominalist, combined with the promise of a full understanding of how nature works, led to a widespread belief that humans can, and should, control nature itself. In this way, humanity could be free

6. Pinckaers, *The Sources of Christian Ethics*, 242.

from nature, able to create whatever kind of world—and ultimately lives—humans desire.

While the result is often called the scientific mind, a more accurate description is the engineering mind-set. Science was different from alchemy in asking "why" rather than "how." Alchemists sought a process that could turn lead to gold, while Johannes Kepler, an early scientist, said: "Oh God, I am thinking Thy thoughts after Thee." In contrast, engineering seeks knowledge to control—it asks what steps to take to achieve an effect. It is not that engineering never asks "why"—instead, engineering always asks "why" in the context of the larger question "how"? Once science lost faith in God, nature lost its purpose, and started converting "why" questions to "how." As Lawrence Krauss says:

> At the same time, in science we have to be particularly cautious about "why" questions. When we ask, "Why?" we usually mean "How?" If we can answer the latter, that generally suffices for our purposes. For example, we might ask: "Why is the Earth 93 million miles from the Sun?" but what we really probably mean is, "How is the Earth 93 million miles from the Sun?" That is, we are interested in what physical processes led to the Earth ending up in its present position. "Why" implicitly suggests purpose, and when we try to understand the solar system in scientific terms, we do not generally ascribe purpose to it. So I am going to assume what this question really means to ask is, "How is there something rather than nothing?" "How" questions are really the only ones we can provide definitive answers to by studying nature, but because this sentence sounds much stranger to the ear, I hope you will forgive me if I sometimes fall into the trap of appearing to discuss the more standard formulation when I am really trying to respond to the more specific "how" question.[7]

By replacing all the "why" questions with "how," science becomes one end of a continuum where knowledge of nature (there is nothing besides nature to know) is sought for its own sake.

At the other end of the continuum, knowledge is sought purely for operational reasons; to be able to drive a car or use an application. Operational knowledge mimics Kepler's desire for knowledge, but rather than thinking the thoughts of God, it seeks to think the thoughts of other men—the engineer who designed a product.

Engineering stands in the middle of this continuum, seeking to solve problems by controlling the natural world, applying knowledge gained

---

7. Krauss, *A Universe from Nothing*, 143–44.

through a process similar to the scientific method. The engineering method is: identify a problem to be solved, hypothesize a set of possible solutions, test each solution, deploy the "best" solution, and examine the real-world results to either improve or replace the solution in the future. Science typically begins with a *piece of knowledge to be acquired* and is "done" when the knowledge is gained. Engineering, on the other hand, starts with a *problem to solve* and is a *forever process* of *continual improvement* and *disruption.*

The engineering mind-set, then, is empirical because it seeks answers through experience, functionally naturalistic and materialistic because it seeks answers primarily in the physical world, and strongly progressive—a topic considered in more detail in the next section. During the Industrial Revolution, people held the engineering mind-set in high regard. In 1910, for instance, George Melville, a retired Rear Admiral of the U.S. Navy, argued his time was the "age of the engineer." Because of this, engineers should lend their "trained ability and judgment" to the perfection of the government, influencing and guiding the "non-expert part of the population."[8] During the Industrial Revolution, according to John Jordan, social problems were seen as managerial problems rather than moral ones, which meant they could be solved by "troubleshooting and problem solving."[9] Jordan says thinkers during this time believed the "same methodology that enabled steel to be manufactured to previously unattainable degrees of hardness could solve ethnic tensions or alleviate poverty."[10]

Technological Progress

Technological progress, the idea that humanity can create a perfect society through the application of technology, is a third crucial intellectual root of neurodigital media. Technological progress makes three essential claims. The first claim is that anything that is new or now is better than whatever has been before. This belief, that the new is always better than the old, is what C. S. Lewis calls "chronological snobbery," which he says is "the uncritical acceptance of the intellectual climate common to our own age and the assumption that whatever has gone out of date is on that account discredited."[11] Rather than accepting chronological snobbery, Lewis argues

---

8. Melville, "The Engineer's Duty as a Citizen," 529.

9. Jordan, *Machine-Age Ideology,* loc. 239.

10. Jordan, *Machine-Age Ideology,* loc. 239.

11. Lewis, *Surprised by Joy,* 207.

You must find why it went out of date. Was it ever refuted (and if so by whom, where, and how conclusively) or did it merely die away as fashions do? If the latter, this tells us nothing about its truth or falsehood. From seeing this, one passes to the realization that our own age is also "a period," and certainly has, like all periods, its own characteristic illusions. They are likeliest to lurk in those widespread assumptions which are so ingrained in the age that no one dares to attack or feels it necessary to defend them.[12]

The second claim is that the clock cannot be—and should not be—turned back. The preference for the new over the old creates a ratcheting effect; once something is invented, *some use* must be found for it.

Third, each new thing has the power to change the fundamental material and ethical norms of human life—everything is a unicorn. J. B. Bury says, "Progress is both an interpretation of history and a philosophy of action,"[13] a "theory which involves a synthesis of the past and a prophecy of the future."[14] According to technological progress, history is the permanent condition of opening an infinite series of Pandora's boxes, each of which must make life better for man, and none of which can be closed.

Bury begins his history of the idea of progress with Augustine's Christian view of history.[15] Christianity's linear view of history has two distinctive features. First, according to Bury, Christianity views history as bound by the authority of God through providence. Truth, therefore, is not something to discover, but rather something accepted because of the authority of God.[16] Second, history is a story of a degrading world that must eventually be rescued by the action of God. Bury says this is because "every child will be born naturally evil and worthy of punishment, a moral advance of humanity to perfection is plainly impossible."[17]

Despite these points, which Bury argues prevented the idea of progress from taking hold of the popular imagination, he does credit Christianity with breaking the cyclical view of time held by the Greeks, which had dominated popular thought before Christianity became widespread. Bury also

12. Lewis, *Surprised by Joy*, 207–8.

13. Bury, *The Idea of Progress*, 1.

14. Bury, *The Idea of Progress*, 3.

15. Bury, *The Idea of Progress*, 21.

16. This reading of Christian theology, however, does not differentiate between the source of truth and the quest for or discovery of truth. God being the source of all truth does not necessarily relegate man to accepting all things "on authority"; there are many places in the Scriptures where God encourages discussion and discovery.

17. Bury, *The Idea of Progress*, 21.

credits Christian theology with proposing a *purpose* for history, a narrative into which all events could fit. It is here the influence of Descartes enters the story once again.

Descartes searched for the proof of God's existence through purely rational means by doubting everything he could doubt, searching for that which he could not doubt, and then rebuilding his entire belief system on that which cannot be questioned. The result was the famous maxim *cogito, ergo sum,* or *I think, therefore I am.*[18] The process Descartes used to come to this point—doubt everything possible, and build up from there—presupposes the only way to true knowledge is through doubt, and faith can only be valid if it can be rationally justified. This foundation of doubt undermined the very concept of truth issuing from authority, one of the significant pillars of Augustinian thought.

The second pillar of Augustinian thought, the idea of God's providence guiding history, was attacked through the immutability of nature, or natural laws. William of Ockham argued that God "could not create universals because to do so would constrain his omnipotence."[19] Ockham's argument sets the providence of God against the immutability of nature—either one is true or the other. Providence requires a God who is in control of nature, who can change nature to accomplish his ends. The study of nature through the scientific method, however, requires nature to be immutable, unchanging, or uniform across time. If nature is immutable, then it seems God is no more able to change nature than is any other force—calling the very idea of providence into question. The success of the methods proposed by Bacon seemed to show that nature is, in fact, immutable, proscribing the providence of God into the narrow sphere of spiritual truth or abandoning the providence of God to altogether.

Further, the success of the new science proposed by Bacon, described in the last section, caused an air of optimism to spread throughout popular culture. Once religious authority and the straitjacket of providence were cast off, there seemed to be no reason humans could not solve any problem they faced by applying the engineering mind-set. Technology progressed at a rapid pace, seemingly validating this general optimism, transferring trust from religious belief to the scientist and engineer—at least through the Roaring Twenties. The first World War—the Great War—shattered the illusion that increasing technological innovation was also improving humankind.

18. Although, rather humorously, Descartes's original line of thinking revolved around his inability to *not doubt,* so more properly he began with *I doubt, therefore I am*—because it is impossible to doubt without there being a person who is doing the doubting.

19. Gillespie, *The Theological Origins of Modernity,* 22.

Rather than being used only to improve the world, technology had, for the first time, been used to destroy on a large—in fact, unprecedented—scale.[20]

In the Great Depression that followed, faith in progress was shaken, but not completely broken. Perhaps the engineering mind-set should be applied to social problems in the same way it was used to solve other issues humans faced. These experiments came to an end with World War II. In this war, the destruction was more significant—including nuclear weapons that could destroy entire cities—and the genocide wrought by the National Socialists and their allies exposed the depth of horror to which the human soul could fall.

The idea of progress appeared to lie in ruins in the aftermath of the World War II. What was needed was a new kind of technology—something completely different in scope and impact than anything that had gone before, a new beginning. Beginning in the 1950s, the digital computing revolution, ecosystem theory, and network theory combined with fresh promise for progress. The movement came to flower in the 1960s, started impacting the broader culture in the 1980s, became ubiquitous through most of the world by the 2000s, and began fading into invisibility as the 2000s progressed. In 1989, when the Berlin Wall fell, creating hope in the transformation of the formerly socialistic countries, the progressive ideal regained its position of power. It applied the engineering mind-set to politics and social structures using the power of digital computing, avoiding war entirely. The nominalism of Descartes morphed into post-structuralism and multiculturalism, creating a world that "moved forward by mutual learning and accommodation rather than war or conquest," creating a future of "productive encounter."[21]

Naturalism

While naturalism is an ancient belief system, according to James Turner, "disbelief in God was literally a cultural impossibility"[22] until cultural secularization began in the early seventeenth century. Turner contends that there were three reasons for this secularization of culture: the secular state stepping into religious disputes to maintain the stability of civil society,[23] economic innovations that moved power out of the church into the hands

20. Gillespie, *The Theological Origins of Modernity*, 7.

21. Gillespie, *The Theological Origins of Modernity*, 10.

22. Turner, *Without God, Without Creed*, 2.

23. Turner, *Without God, Without Creed*, 9.

of newly forming secular elites,[24] and a new approach to knowing physical and human reality.[25]

The first two of Turner's elements are about authority, which is directly connected to the religious beliefs of a culture by determining who counts as the ultimate source of truth and who is allowed to create and shape the education of new citizens.[26] The rising tide of personal autonomy undermined all authority—including God's authority over individuals—placing authority for how each person lives their lives in their own hands. The idea of progress is, in a sense, also a form of autonomy—taking the future of humanity out of the realm of God's providence and placing it in human hands.

Turner's third element is the engineering mind-set; if man can discover reality without God, then what need is there of God? According to Lawrence Krauss:

> This reflects a very important fact. When it comes to understanding how our universe evolves, religion and theology have been at best irrelevant. They often muddy the waters, for example, by focusing on questions of nothingness without providing any definition of the term based on empirical evidence. While we do not yet fully understand the origin of our universe, there is no reason to expect things to change in this regard. Moreover, I expect that ultimately the same will be true for our understanding of areas that religion now considers its own territory, such as human morality.[27]

Some Christian thinkers reacted to this secularization of culture by separating the natural and supernatural, and sources of knowledge about nature from sources of knowledge about God.[28] Turner says: "Perhaps man could know God through inner light and divine revelation as well as empirical rationality. But he could learn of nature and history in only one thoroughly reliable way: the new way."[29] This new way of knowing nature excluded asking why because "why" is a query with no content. "How" is the only question worth asking.

As with the other movements considered above, naturalism played a significant role in the culture of Silicon Valley, where neurodigital media developed. Fred Turner says the pioneers of the computing revolution

24. Turner, *Without God, Without Creed*, 10.

25. Turner, *Without God, Without Creed*, 13–14.

26. Turner, *Without God, Without Creed*, 12.

27. Krauss, *A Universe from Nothing*, xv.

28. Turner, *Without God, Without Creed*, 37.

29. Turner, *Without God, Without Creed*, 29.

intentionally developed an "alternate temple" of the "church of technology," explicitly set against Christianity;[30] Richard Barbrook and Andy Cameron call it a "naturalizing" ideology.[31] Naturalism continues to be the default belief system in Silicon Valley. Caroline McCarthy says: "At Google, few co-workers would blink an eye if you told them that you spent the previous weekend attending an electronic music festival in an otter costume, but you might get some funny looks if you admitted you went to church every weekend."[32]

Hippies, Cold Warriors, and the Californian Ideology

These four—personal autonomy, the engineering mind-set, the idea of progress, and naturalism—came together in the 1950s and 60s to form what Richard Barbrook and Andy Cameron call the Californian Ideology, "offer[ing] a fatalistic vision of the natural and inevitable triumph of the hi-tech free market."[33] Californian Ideology grew out of the collision and merging of two communities that were markedly different—at least to outside observers—the hippies and the cold warriors.

Hippies

The Beats of the 1950s inspired the hippies of the 1960s. The Beats were a literary subculture that was, according to John Anthony Moretta, "apolitical and fatalistic as a result of living with the prospect of nuclear holocaust."[34] The Beats held that the "American political system had become so corrupt and controlled by corporations that it was beyond any sort of redemption."[35] As the movement faded in the late 1950s, the few remaining Beats moved into the Haight-Ashbury area of San Francisco because of the abundance of low-priced large Victorian houses well suited to communitarian living, forming a countercultural core. By the mid-1960s, and the beginning of the war in Vietnam, younger people who decided they could no longer support the middle-class culture, which they considered suffocating, flocked to the Haight-Asbury area to "drop out" and "do your own thing." These hippies

30. Turner, *From Counterculture to Cyberculture*, loc. 1170.
31. Barbrook and Cameron, "The Californian Ideology."
32. McCarthy, "Silicon Valley Has a Problem with Conservatives."
33. Barbrook and Cameron, "The Californian Ideology."
34. Moretta, *The Hippies*, 14.
35. Moretta, *The Hippies*, 14.

were "anti-traditional values, anti-consumption, and anti-materialism," participating in a "Great Refusal."[36]

By the end of the "Summer of Love" in 1967, the Haight-Ashbury area had become a den of drugs and sexual crime; a place that was "supposed to float on love" was "charged with strife."[37] Throughout the 1960s, as individual hippies burned out on the scene in San Francisco, they moved out into the countryside, forming communes. While the communes were communities, they also placed the individual first—and they were places where people could go to commune with nature. The commune movement was both deeply progressive and regressive at the same time, seeking a backward-looking, individualistic, nature-based life along the lines of Thoreau's *Walden* by taking advantage of modern technology and newly developed ways of seeing ecosystems and organizations.

Cold Warriors

In 1945, at the end of World War II, the allies who had formerly fought against the Axis states, including Germany and Japan, entered into a war without directly engaging in military operations—the Cold War. This Cold War lasted from 1945 until 1991, when the Soviet Union dissolved into a group of member states. Many regional wars, such as those in Korea and Vietnam, were fought as proxies between the opposing sides. Throughout this time, the military doctrine of mutually assured destruction was followed; enough nuclear weapons were deployed and maintained to enable the destruction of any nation who struck first. The maintenance of parity, however, required constant innovation in technology and tactics. If either country fell behind technologically, for instance because its launch sites could be discovered and destroyed, the result would be complete dominance.

The United States government quickly repurposed the research centers, which had helped win World War II in 1945, at the beginning of the Cold War. While much of the nuclear and weapons research was done in Massachusetts, New Mexico, and Tennessee, research into digital computing was centered in Los Angeles and Massachusetts. With money provided by the government and large corporations, the University of California in Los Angeles (UCLA) built a computer science program unrivaled in the world. It was at UCLA that a pair of young students, Steve Crocker and Vint Cerf, stole into the computer lab to gain extra time on the machines located there. A small team of graduate students at UCLA in the late 1960s,

36. Moretta, *The Hippies*, 37.
37. Moretta, *The Hippies*, 184.

including Steve Crocker, Bob Kahn, and Vint Cerf, invented the protocols that underlie the ARPANET, leading to the creation of the global Internet.

Over the next forty years, Xerox opened its famed Palo Alto Research Center, where the mouse and graphical user interface were invented; Intel, a maker of computer processors, was founded in Santa Clara; Hewlett-Packard was founded in Palo Alto; Apple Computers was founded in Los Altos; Cisco Systems, a maker of computer networking software and hardware, was founded in San Francisco; and Facebook was founded in Menlo Park. While the area around the Massachusetts Institute of Technology (MIT) continued to play a significant role in the development of computing and communications technology, Los Angeles, and later in Silicon Valley, a collection of cities just south of San Francisco, were the heart of the digital revolution.

The researchers working in these labs had a culture distinct from many other research and academic cultures throughout the world. In this culture, a sense of community and open lines of communication were combined with overriding personal autonomy and naturalistic presuppositions.

The Californian Ideology

The research communities around Los Angeles and Silicon Valley did not, from the outside, seem to have much in common with the hippies, their drugs and communes, situated in the same area. One dedicated itself to helping win a war; the other committed itself to ending all wars. One focused on building new technologies used to improve the modern lifestyle; the other focused on escaping that very lifestyle and living close to the land. But these two communities, however different their goals might have been, shared a common culture—curious, playful, communal and yet individualistic, and possessing a strong aversion to authority.

The hippie movement was communally and individualistic, focused on reducing the power of authorities of all kinds, both religious and secular, and giving power to individuals so they could shape their lives in any way they liked. Similarly, Barbrook and Cameron say, "Information technologies . . . empower the individual, enhance personal freedom, and radically reduce the power of the nation-state. Existing social, political, and legal power structures will wither away to be replaced by unfettered interactions between autonomous individuals and their software."[38]

The hippie movement held to strong naturalism; people should return to—and live close to—nature. Drugs, such as LSD, were a natural and

---

38. Barbrook and Cameron, "The Californian Ideology."

effective way to bring about spiritual states without needing any real gods or goddesses. Fred Turner maintains the founders of the computing revolution "began to imagine institutions as living organisms, social networks as webs of information, and the gathering and interpretation of information as keys to understanding not only the technical but also the natural and social worlds," as they "turned away from political action and toward technology and the transformation of consciousness as the primary sources of social change." [39] William Barrett claims the tendency towards a naturalistic and mechanistic vision of reality is irresistible because of the very nature of the digital computer itself, which "seems able to reproduce the process of the human mind." [40]

Finally, the hippie movement was progressive, believing the world could be perfected if only the right combination of social and personal beliefs and systems could be promoted and employed—such as the communal farming model. Hippies also believed technology must play a role in the emancipation of humans by improving agricultural methods, living arrangements, and the kinds of drugs needed to produce spiritual experiences. The cold warriors were progressive in believing they could create a peaceful world, and then change the world itself, through technology.

While the goals might be different, the cultures and values were often the same. Stewart Brand brought the hippies and cold warriors together through a series of what Fred Turner calls network forums, physical meetings augmented with the *Whole Earth Catalog*.[41] The result was an explosion of new ideas, the creation of the digital computing world, and eventually, the creation of neurodigital media.

39. Turner, *From Counterculture to Cyberculture*, loc. 110.
40. Barrett, *Death of the Soul*, 154.
41. Turner, *From Counterculture to Cyberculture*, loc. 1069.

3

The Person

Neurodigital media impacts the *person*—the individual human being—in part because it is built on a particular view of the person. How is each person understood within the Californian Ideology? How are they treated as individuals and within groups? Because the Californian Ideology is naturalistic, it tends to see the person as a product of natural forces and a part of the natural world. Because the Californian Ideology is progressive, it tends to see the person—and groups of people—as improvable. By shaping people, individually and in groups, the world can be brought ever closer to an ideal state.

These are questions of anthropology, what a worldview believes about human beings. This chapter begins with a short look at naturalism, and then explores naturalistic anthropology. The second part of the chapter explores a broadly theistic view of the person; although references primarily from within Christianity will be used in this section, theists of all kinds will find common ground with the picture of the person presented. These two views are then compared in three key areas: the dignity of the person, the freedom of the person, and the social aspects of the person.

Naturalism and the Person

Worldviews and belief systems are not monolithic; while every person who holds to a worldview will hold some set of overlapping beliefs, they will not all hold identical beliefs. The extremes in Christianity vary from those who wade into deep theological waters to those who just attend church once a

38

year (if ever) and are personally uncertain about many Christian beliefs—but all Christians believe that Jesus of Nazareth was crucified and raised from the dead to cover the sins of the faithful. Jewish beliefs vary from being largely a culture to deep study of the Torah, but all Jewish believers hold the exodus from Egypt through the Passover by the hand of God is the origin of the Jewish nation.

Naturalism and atheism are wide-ranging belief systems, as well, including fatalistic, logical positivistic, and largely cultural branches—but all naturalists and atheists share a strong anti-theism. C. S. Lewis says this anti-theism holds the "ultimate Fact, the thing you can't go behind, is a vast process in space and time which is going on of its own accord."[1] According to Carl Sagan, "The Cosmos is all that is or was or ever will be. Our feeblest contemplations of the Cosmos stir us—there is a tingling in the spine, a catch in the voice, a faint sensation, as if a distant memory, of falling from a height. We know we are approaching the greatest of mysteries."[2]

A Lack of Meaning

For naturalism, because there is no external "Fact," as Lewis says, there is also no external *meaning* or *telos* within the universe or imposed from without. While it is possible to ask *why* something is in terms of process, it is not possible to ask *why* something is *intended*—there is no external "Fact" that can impart meaning. Returning to Kraus, *why* essentially becomes an equivalent of *how,* and *why* becomes a nonsensical question.[3] Or, in the words of Ken Boa, naturalism "gives no meaning for the particulars of nature or mankind."[4] The universe—and life itself—are reduced to their mechanisms; William Barrett says "the science of mechanics was no sooner founded than a widespread ideology of mechanism followed in its wake."[5]

If the world has no *why,* no *telos,* and human beings are *merely* another product of long, random process, wholly contained in the world, then it is clear human life has no meaning that reaches beyond nature, either. Whatever there is for humans to pursue, it cannot be beyond nature because *nothing exists beyond the natural world.* Any value human beings possess must be found within the world and expressed in purely materialistic terms.

1. Lewis, *Miracles,* 7.
2. Sagan, *Cosmos,* 1.
3. Krauss, *A Universe from Nothing,* 143–44.
4. Boa, "What Is Behind Morality?," 17.
5. Barrett, *Death of the Soul,* xv.

The soul, being something outside the physical universe, is also banished in naturalism. John Barresi and Raymond Martin say: "In sum, whereas throughout most of the seventeenth century the self had been a soul, by the end of the eighteenth century it had become a mind, albeit one whose status as a real entity was obscure. One mystery, the immaterial soul, had been dropped. Another, the self as material mind, had emerged to take its place."[6] Nicolas Carr traces the death of the soul through Descartes's dividing of the mind from the soul. At first, this division seemed to provide comfort for religious believers, as it left the mind and soul free from the observations of science. Over time, however, as "reason become the new religion of the Enlightenment, the notion of an immaterial mind lying outside the reach of observation and experiment seemed increasingly tenuous."[7]

Joel Green, in his study on the nature of humanity, says, "Reductive Materialism has it that the human person is a physical (or material) organism, whose emotional, moral, and religious experiences will ultimately and decisively be explained by the natural sciences. People are nothing but the product of organic chemistry."[8] Carr says, "Scientists rejected the 'mind' . . . even as they embraced Descartes' idea of the brain as a machine." Thoughts, memory, and emotion came to be seen as outputs of the physical outputs of the brain.[9] Ultimately any activity by the person must be explainable in purely observable terms—including human freedom, human dignity, and the formation of relationships and community.

The Mind as Machine

Once the person is reduced to the physical, the mind can be reduced to a computer—an object to be acted on and "reprogrammed" using processes applied through neurodigital media. A. M. Turing, a significant figure in the development of machine computing, describes the mind in just these terms. After describing the operations a computer must go through to solve a problem, he states: "We may now construct a machine to do the work of this computer."[10] At the time Turing wrote this, *people* were computers, rather than machines,[11] so Turing's argument implies a computing machine can perfectly replace human computers. In a later paper, Turing directly

6. Barresi and Martin, *Naturalization of the Soul*, 1.

7. Carr, *The Shallows*, loc. 424.

8. Green, *Body, Soul, and Human Life*, 30.

9. Carr, *The Shallows*, loc. 424.

10. Turing, "On Computable Numbers," 251.

11. Light, "When Computers Were Women."

equates human intelligence with machine intelligence, proposing a game where a person tries to determine if a conversation partner is a machine or another person.[12]

Norbert Weiner takes Turing's test, and the later Chinese Room, to its logical conclusion, equating communication with a machine to communicating with another person—"the fact that the signal in its intermediate stages has gone through a machine rather than through a person is irrelevant and does not in any case greatly change my relation to the signal."[13] Marcin Miłkowski makes the connection the mind and computers explicit, saying "brains are computers, or, to be precise, information processing mechanisms."[14]

If computing machines can replace humans, and all intelligence is computational, then humans are instruments in the same way computing machines are—individual users, as nodes within the social graph built by using neurodigital media, have no intrinsic value, and can be shaped to better fit into the overall purpose of the system. Brent Waters, for instance, observes that once the computational theory of mind is accepted, "biological embodiment is an accident of history," so "augmenting or replacing the body is a more efficient continuation" of evolution.[15]

Theism and the Person

Christianity, in stark contrast to naturalism, begins with the supernatural. God existed before the natural world, creating the natural world as described in the first verse of Genesis: "In the beginning, God created the heavens and the earth." Because a personal being created the universe, it was created for a *purpose;* the world has a target beyond itself—God. As Psalm 19:1 says, "The heavens declare the glory of God, and the sky above proclaims his handiwork."

Humans, individually and corporately, as a part of God's creation, also have a purpose, a target beyond the natural world. Wayne Grudem cites Isaiah 43:7, "everyone who is called by my name, whom I created for my glory, whom I formed and made," to argue the human race was created for

---

12. Turing, "Computing Machinery and Intelligence."

13. Wiener, *The Human Use of Human Beings*, 16.

14. Miłkowski, "Why Think That the Brain Is Not a Computer?," 22.

15. Waters, *From Human to Posthuman*, 42.

the glory of God.[16] C. S. Lewis adds communion with God (or being in community with God) as a purpose for the creation of humans.[17]

Humans point beyond the natural universe in terms of their creation as well as their purpose. In Christian anthropology, humans are created by God in his image—the *imago Dei*—as described in Genesis 1:26, "Let us make man in our image, after our likeness." Many early Christian scholars held the *imago Dei* was the "spiritual" part of the person, while later scholars have held the *imago Dei* is primarily physical in form.[18] According to K. A. Mathews, however, neither of these views are "at odds with Hebrew anthropology."[19] Rather than being either physical or spiritual, the *imago Dei* represents the whole person in *psychosomatic union*. Genesis 2:7 depicts this union, saying "the Lord God formed the man of dust from the ground and breathed into his nostrils the breath of life, and the man became a living creature"—the body and spirit are *both* required to make the whole person. Charles Ryrie says "man was created a total being, material and immaterial, and that total being was created in the image of God."[20]

One of the baffling things about the *imago Dei* is that the Scriptures *never define what being made in the image of God means*. This has opened the door for endless speculation throughout history, with a few theories rising to the top of the theoretical pile. Calvin held the *imago Dei* resided in the intelligence of man, and Kant that the image was in the I/thou relationship. An alternative to each of these is the image represents all of these things. Adam and Eve were created in the image and likeness of God in the same way Seth was born in the likeness and image of Adam (Genesis 5:7). Wayne Grudem says the text "does not specify any specific number of ways that Seth was like Adam, and it would be overly restrictive for us to assert that one or another characteristic determined the way in which Seth was in Adam's image and likeness."[21]

Even though there is no way to catalog a "definitive list" of the ways in which humans are made in the image of God—again, the Scriptures do not give such a list—there are at least three things that can definitely be said to be "in the list." The *imago Dei* means humans have moral freedom and responsibility. The *imago Dei* means at least that humans have a fundamental dignity no other being or thing in the created order has. The *imago*

---

16. Grudem, *Systematic Theology*, 440.

17. Brazier, *C. S. Lewis*, 28.

18. Mathews, *Genesis 1–11:26*, 165–66.

19. Mathews, *Genesis 1–11:26*, 167.

20. Ryrie, *Basic Theology*, 219.

21. Grudem, *Systematic Theology*, 444.

Dei means humans were created in and for relationship with others—God, individually with other humans, and in a larger community.

Freedom and Responsibility

Perhaps one of the most remarkable things about the United States Declaration of Independence is not that it signaled the formation of a new country, but rather that it paid deep homage to the rights, freedoms, and responsibilities (a point often forgotten in the modern world) of individual citizens. The Declaration holds "these truths to be self-evident, that all men are created equal, that they are endowed by their Creator with certain unalienable Rights, that among these are Life, Liberty and the pursuit of Happiness." The idea of individual freedom, however, has not been a feature of most cultures throughout time.

Theism and Freedom

Ibn Rush'd (often Latinized as *Averroes)* held to a form of panpsychism or unicity;[22] there is one mind in the universe that animates all creatures.[23] Ibn Rush'd was widely read and respected through the time of William of Ockham; many Christian scholars accept panpsychism. For instance, Signer of Brabant, in 1270, said there was a single "separately existing intellect . . . common to all humans."[24] Orthodox Christian scholars, however, rejected panpsychism because it undermined the moral freedom of individuals. Thomas Aquinas, one of the most influential theologians in Christian history, argued that "our idea of God and our relationship with him depends largely on our concept of freedom."[25]

Freedom and responsibility are tightly tied in the Scriptures,[26] such as setting aside sanctuary cities. In ancient Israel, a family would designate an avenger of blood on the death of a family member. The avenger would seek out and kill the person responsible for the death as required by the law—"life for life." In Deuteronomy 19:1–3, God instructs Israel to set aside three cities of refuge in which a killer can seek protection from the avenger

22. Hasse, "Influence of Arabic and Islamic Philosophy."
23. Geisler, "Averroes," 63.
24. McInerny and Thomas (Aquinas), *Aquinas Against the Averroists*, 10.
25. Pinckaers, *The Sources of Christian Ethics*, 327.
26. There are two broad concepts of freedom in Christian theology—libertarian freedom and compatibilistic freedom. The argument made here does not assume either view is correct, only that some form of human freedom and moral responsibility is true.

of blood. So long as the killer did not leave the city proper, and in some cases the general area of the city, they were safe from the avenger of blood. These cities were spaced within Israel so they could be reached within a few days of travel on foot, giving everyone an equal chance to seek protection, and God gave instructions to expand the number of cities available if Israel became a geographically larger nation.

Not every killer could seek the protection of a sanctuary city, however—only manslayers. What is a manslayer? According to Deuteronomy 19:4, it is someone who "kills his neighbor unintentionally without having hated him in the past." The next verse provides a helpful example, "as when someone goes into the forest with his neighbor to cut wood, and his hand swings the axe to cut down a tree, and the head slips from the handle and strikes his neighbor so that he dies." The difference between murder and manslaughter is the *intent* of the killer, and the killer's responsibility is different in each case. In one case, the killer has chosen to kill, and the penalty is death. In the other, the killer's actions were unintentional, and the punishment is fleeing to a city of sanctuary for some time. Another instance of this differentiation is found in the laws about how to handle an ox that gores a person—whether or not the owner is responsible depends on whether the ox was known to be violent in the past.

These are not the only cases where the freedom to make moral choices is assumed in the Scriptures. In Joshua 24:15, Israel is given the choice of serving God or not—"choose this day whom you will serve." In 2 Corinthians 5:20, Paul urges his readers to be "reconciled to Christ." In both of these passages, the Scriptures assume individuals have the freedom to choose whether or not they will follow God. As Augustus Hopkins Strong says, "man, when he appears upon the scene, is no longer brute, but a self-conscious and self-determining being, made in the image of his Creator and capable of free moral decision between good and evil."[27]

Naturalism and Freedom

Naturalism presents a far more interesting—and in some ways conflicted—view of human freedom. The most consistent form of naturalism holds that because every decision a person makes, and every emotion a person feels, is a product of purely physical processes, there is no such thing as free will. As C. S. Lewis says:

---

27. Strong, *Systematic Theology*, 472.

[N]o thoroughgoing Naturalist believes in free will: for free will would mean that human beings have the power of independent action, the power of doing something more or other than what was involved by the total series of events. And any such separate power of originating events is what the Naturalist denies. Spontaneity, originality, action "on its own", is a privilege reserved for "the whole show", which he calls Nature.[28]

Tom Wolfe agrees: "Since consciousness and thought are entirely physical products of your brain and nervous system—and since your brain arrived fully imprinted at birth—what makes you think you have free will?"[29] Wolfe declares the result is the reduction of the moral decision-making ability of human beings to chemical processes—"The fix is in! We're all hardwired! That, and: Don't blame me! I'm wired wrong!"[30] In the words of William Barrett, "as our molecules go, so do we."[31]

Christian thinkers are not the only ones who hold that naturalism inevitably leads to determinism. Carr, for instance, argues that when the soul was suppressed through the scientific mind-set, "memory, and emotion, rather than being the emanations of a spirit world, came to be seen as the logical and predetermined outputs of the physical operations of the brain. Consciousness was simply a by-product of those operations"[32] Note the word *predetermined* in Carr's description.

This consistently naturalistic belief in determinism is set off within Californian Ideology by a strong belief in individualistic, personal—nihilistic—freedom. One of the definitional essays of the Ideology giving rise to neurodigital media describes the goal as a world where "[e]xisting social, political, an legal power structures will wither away to be replaced by unfettered interactions between autonomous individuals and their software."[33] John Perry Barlow, in his seminal essay from 1996, states:

Cyberspace consists of transactions, relationships, and thought itself, arrayed like a standing wave in the web of our communications. Ours is a world that is both everywhere and nowhere, but it is not where bodies live. We are creating a world that all may enter without privilege or prejudice accorded by race, economic power, military force, or station of birth. . . . Your legal concepts

28. Lewis, *Miracles*, 7.
29. Wolfe, "Sorry, But Your Soul Just Died."
30. Wolfe, "Sorry, But Your Soul Just Died.".
31. Barrett, *Death of the Soul*, xv.
32. Carr, *The Shallows*, loc. 474.
33. Barbrook and Cameron, "The Californian Ideology."

of property, expression, identity, movement, and context do not apply to us. They are all based on matter, and there is no matter here. Our identities have no bodies, so, unlike you, we cannot obtain order by physical coercion.[34]

Others argued, in the same time frame, that the "Internet would soon dissolve the bureaucracies of the marketplace by stripping away the material bodies of individuals and corporations."[35]

It seems, in this view, that once the conscious person emerges from physical processes, they can be entirely freed from those physical processes. The mind, once destroyed on the altar of lack of scientific evidence, has not only returned, but slipped its physical moorings to sail out onto a digital sea of absolute freedom.

"The person is to stand at the center of the technological world, using technology to liberate us from the chores that prevent us from having the leisure time we need to be human."[36] The person is made free to be truly human by technology, freed even from making difficult decisions. And yet, at the same time, according to Richard Thaler and Cass Sunstein, technology is to be used to "self-consciously [attempt] to move people in directions that will make their lives better."[37] The person is to be perfectly free, and yet also to be nudged towards a destination that is "better for them"—all the while ignoring what the meaning of the word *better* might be.

The naturalistic view of the person, then, is both mechanistic and emergent, both strictly tied to the physical and yet able to abandon the physical to rebuild itself in a digital world without organizations and laws. The person is deterministic, doing what the underlying molecules say to do, and yet fully able to decide what is best, and yet again needing guidance in making choices that are "best for them." In Thaler and Sunstein's view, the need is for a libertarian paternalism, somehow treating the person as an instrumental and shapable end with no freedom, and yet respecting the person as making decisions free of all outside influence.[38]

There is no attempt to resolve these contradictions in the naturalistic anthropology underlying neurodigital media. Sometimes absolute freedom will come to the fore, while managing the user to some specified end will be point of focus at other times. Those who build, operate, and profit off systems built on neurodigital media will frame their systems in terms of

34. Barlow, "A Declaration of the Independence of Cyberspace."
35. Turner, *From Counterculture to Cyberculture*, loc. 232.
36. Gertz, *Nihilism and Technology*, loc. 125.
37. Thaler and Sunstein, *Nudge*, 10.
38. Thaler and Sunstein, *Nudge*, 10.

absolute freedom of choice, while designing the systems to restrict freedom to "provide a service," or "do what is better for the user." Both elements inter-mix freely, but almost always in ways that ultimately guide decisions down a path that brings the user more closely to the worldwide of the builder, or extracts profit from the user for the operator.

Dignity in Contrast

Human dignity is difficult to define, but it is generally taken to mean that each person has value that is not tied to their mental or physical capabili-ties, and is not diminished because of age (either young or old) nor by the commission of a crime. For instance, the Constitution of the United States forbids cruel and unusual punishment on the grounds that even criminals have essential human dignity. Dignity is tied to freedom as well as value; Kant says, "Autonomy, then, is the reason for the dignity of Human and all reasonable natures."[39] Humans have individual worth and dignity in both Christian and naturalistic anthropologies—but the reasons for this dignity, and the resulting view of the individual person, are very different.

Dignity in Christianity

Dignity in Christian anthropology is sourced directly from the *imago Dei*. This tie is made explicit through Genesis 9:6, where God inveighs against murder by saying: "Whoever sheds the blood of man, by man shall his blood be shed, God made man in his own image." Within Christian anthropology, the value of the person is *intrinsic* because it is a part of the essence of the human being, and it is does not depend on anything external to the person determining their value.

The Christian view of the value of individuals can be traced back through its Jewish roots. Many ancient religions held the person had very little value because only members of the royal family—specifically the king or princes—were considered the sons of god.[40] According to J. Richard Middleton, however, Judaism decisively shifted this view by holding every person is a son of God.[41] Nahum Sarna illustrates this point through the sal-utations given of the kings of Mesopotamia: "The father of my lord the king

39. Author's translation of "Autonomie ist also der Grund der Würde der men-schlichen und jeder vernünftigen Natur," in Kant and Kirchmann, *Grundlegung zur metaphysik der sitten*, 63.

40. Novak, "The Judeo-Christian Foundation of Human Dignity," 109.

41. Middleton, *The Liberating Image*, 206.

is the very image of Bel (⬚alam bel) and the king, my lord, is the very image of Bel," and the name of the Pharaoh in Egypt, Tutankhamen, which means: "[T]he living image of (the god) Amun."[42] K. A. Mathews holds that Genesis 5:3 supports human sonship to God as father,[43] saying: "In the ancient Near East royal persons were considered the sons of the gods or representatives of the gods (cf. 2 Sam 7:13–16; Ps 2:7)." Mathews continues, "mankind is appointed as God's royal representatives (i.e., sonship) to rule the earth in his place."[44] According to Dale Patrick[45] and Paul Kissling,[46] applying the image of God to every person raises all of humanity to a position of transcendence, having value far above any other created thing or being within nature.

Michael Novak argues that the apostles follow the Tanakh in insisting "every single human is loved by the Creator, made in His image."[47] As an example, consider Romans 16:22, where the scribe who either aided in the writing of or transcribed Paul's letter to the Romans attaches a personal greeting to those who would be receiving the letter: "I Tertius, who wrote this letter, greet you in the Lord." This passage is significant in understanding the value Christians placed on individual humans because *Tertius* means *third*. James Montgomery Boice notes that Roman slaveholders would name their slaves in order of their birth, so Tertius was likely the third slave born in the household of his owner.[48] Instead of being treated as a slave, however, he was "in the Lord," and hence seen as enough of a person of value within the church to include a separate greeting.

Finally, dignity in Christian anthropology dictates that no person is to be treated *merely* as an instrument to be used to reach some end; as Gordon Lewis and Bruce Demarest say: "Persons as spiritual beings are not things to be folded, mutilated, or spindled. As self-transcendent spirits humans are self-conscious and self-determining subjects and moral agents. Christians see in all other persons active beings who should be free from coercion to think, feel, will, and relate."[49]

---

42. Sarna, *Genesis*, 1989, 12.

43. Mathews, *Genesis 1–11:26*, 170.

44. Mathews, *Genesis 1–11:26*, 164.

45. Patrick, "Studying Biblical Law as a Humanities," 40.

46. Kissling, *Genesis*, 125.

47. Novak, "The Judeo-Christian Foundation of Human Dignity," 109.

48. Boice, *Romans,*1954.

49. Lewis and Demarest, *Integrative Theology*, 172.

Dignity in Naturalistic Anthropology

Secular legal and political theorists hold human dignity in high regard; Jasper Doomen says, "Human dignity serves as a guiding principle for international legislation, and varieties of it lie at the root of a number of constitutions."[50] The connection between personal dignity and the worth of a person has led to a widespread concern with dignity; Remy Debes says it is "obvious that human dignity is a persistent concern of a vast array of cultures,"[51] finding expression in "religion, philosophy, literature, and art of all societies, modern and ancient."[52]

Because naturalism lacks an external source such as the *imago Dei,* there is little agreement about what dignity means. Debes declares that the meaning of dignity is "fraught with confusion, contradiction, disagreement, and a shocking willingness to decide who counts as 'human.'"[53] Leslie Henry surveys the meaning of dignity across many cultures, and holds that the results show there is no "positivistic claim to dignity's meaning." [54] Some political theorists have tried to ground the dignity of individuals in the concept of rights, but Michael Meyer argues rights are insufficient support for dignity, saying: "While having and exercising certain rights is important to our dignity as human beings, what we commonly regard as essential to human dignity would not be explained even if we were able to delineate all of the relevant rights and the particular ways in which each of them expresses or protects human dignity."[55]

Meyer replaces right with freedom as the foundation of dignity; the "person who has human dignity is fundamentally a person who is self-possessed; he at least has the capacity to give direction to his own life."[56] Jeremy Waldron agrees with this assessment, saying:

> Dignity is the status of a person predicated on the fact that she is recognised as having the ability to control and regulate her actions in accordance with her own apprehension of norms and reasons that apply to her; it assumes she is capable of giving and entitled to give an account of herself (and of the way in which she is regulating her actions and organising her life), an account

50. Doomen, "Beyond Dignity," 57.
51. Debes, "Dignity's Gauntlet," 51.
52. Debes, "Dignity's Gauntlet," 51.
53. Debes, "Dignity's Gauntlet," 51–52.
54. Henry, "The Jurisprudence of Dignity," 177.
55. Meyer, "Dignity, Rights, and Self-Control," 521.
56. Meyer, "Dignity, Rights, and Self-Control," 533.

that others are to pay attention to; and it means finally that she has the wherewithal to demand that her agency and her presence among us as a human being be taken seriously and accommodated in the lives of others, in others' attitudes and actions towards her, and in social life generally.[57]

Ron Highfield notes that in the modern view, "[t]o treat humans according to their dignity is to treat them as if they owned themselves and had a right to determine their own actions."[58]

Both definitions of dignity—moral freedom and rights—rely on a form of turning an *is* into an *ought*. If rights exist, it is because they either evolved directly in some way, or they emerged from the evolutionary process, and hence are not fixed in space or time. New ways of interacting with the world and other human beings can cause modifications in what is optimal for survival, shaping the evolutionary process, and hence shaping the meaning of rights and moral freedom. Both definitions of dignity are also problematic in the face of variable human capacity. If moral freedom is taken as the sole source of dignity, then those who are too young to be considered adults, slaves, the addicted, or those otherwise restricted from making moral decisions, are less dignified than those who have fuller forms of moral freedom. If dignity is taken to be a matter of rights, then the imprisoned have less dignity than those with a fuller freedom.

Dignity in naturalistic anthropology is variable, rather than fixed—it depends on the situation. Overriding a person's free choice may increase their dignity in some situations—particularly given naturalism's compromised view of freedom. Overriding a person's moral decision-making is also permitted when it improves their dignity in some way. These shifting roles provide the cover necessary to override individual moral choice and moral freedom when some greater commercial or social good is present—especially the greater good of progress, described later in this chapter.

Relationships in Contrast

The desire for personal relationships, according to R. F. Baumeister and M. R. Leary, "may well be one of the most far-reaching and integrative constructs currently available to understand human nature."[59] After surveying over 250 studies, they conclude "a lack of relationships has ill effects on health and

57. Waldron, "How Law Protects Dignity," 202.
58. Highfield, *God, Freedom and Human Dignity*, loc. 1009.
59. Baumeister and Leary, "The Need to Belong," 522.

social adjustment, the desire for relationships elicits goal-oriented behavior, and the desire for relationships is universal among all persons."[60] However, the foundation of relationships is far different in Christian belief and naturalism. Christianity grounds relationships in the dignity of others and the *imago Dei,* while naturalism grounds relationships in instrumental gain, either for the community, the species, or the individual.

Relationships in Christianity and Theism

For Karl Barth, the male-female relationship of Adam and Eve emphasizes the confrontational I/thou nature of relationships in Christian theology. Since God created man with a companion made from within himself, according to Genesis 2:18–25, God always intended human beings to be "in relationship."

The nature of relationship in Christian anthropology is always "other-focused" or altruistic; each person in the relationship works for the good of the other. As Anthony Hoekema says, "The image [of God] must be seen in man's threefold relationship: toward God, toward others, and toward nature."[61] *Towards* is a key point in all three relationships. A parable often used as an example of the other-focused view of relationships is the story of the long spoons, originally attributed to Rabbi Haim of Romshishok, but repeated in many different contexts. In this story, a man is taken to a room where he observes a group of people sitting around a table. Each person has a bowl full of some soup in front of them, and each one has a spoon that is longer than their arms. Each person is trying, unsuccessfully, to feed themselves with these long spoons. Since the spoons are longer than their arms, however, they cannot—and hence they are miserable and hungry. The man is then taken to another room which is, in all respects, identical to the first room—only the people sitting around the table are feeding one another with these long spoons.

A second point Christian anthropology makes about relationships is that they must be formed within an objective moral standard—relationships have rules that must be followed to be correctly formed. The importance of correctly formed relationships is described in Scripture from the very beginning; Genesis 3:16 describes the consequences of the fall (Adam and Eve stepping outside the bounds of the rules on which their relationship with God is based) in terms of Eve's relationships. Her relationship with her children will now begin in pain, and her relationship with her husband will

60. Baumeister and Leary, "The Need to Belong," 498.
61. Hoekema, *Created in God's Image,* 95.

now be marred, according to Mathews, by a "struggle for mastery between the sexes."[62]

Another instance can be found in Genesis 4:12, where God punishes Cain for murdering Abel by making him a "fugitive and wanderer on the Earth." Cain does not dispute that he has broken the relational rules by murdering his brother, but rather says this punishment is "too much to bear" because he must live his life outside the relationships with his family and God. According to Mathews, the only protection Cain has against being killed by anyone who finds him is God, and God is withdrawing his protection.[63]

Relationships in Naturalism

Naturalistic anthropologies do not generally focus on relationships in and of themselves, but rather on the evolution of altruism, cooperation, and ethical systems that make relationships possible. The evolution of these traits is a functional matter, increasing the ability of individuals or species to survive—hence all naturalistic theories hold to a functional view of relationships. As Eric Alden Smith says:

> Ever since Darwin, the "why" posits an evolutionary design force, rather than a supernatural or teleological one. Thus, ultimate explanations are also known as functional accounts and ask how a trait of interest contributes to adaptive success or inclusive fitness or otherwise is generated or maintained by evolutionary forces such as natural selection and transmission dynamics.[64]

Because no evolutionary record exists of the origins of relationships, and no experiment can recapture the state of humans before relationships existed to examine how they would evolve, there is no definitive explanation for the evolution of the human traits that make relationships possible. Several theories have been offered, however.

As an example, Dennis L. Krebs begins with Darwin's argument that "animals may obtain benefits from group living by exchanging goods and services and by coordinating their efforts."[65] These benefits, however, are not enough to move beyond basic group seeking and towards an apparently objective moral code and altruistic behavior. Stronger individuals can

---

62. Mathews, *Genesis 1–11:26*, 251.
63. Mathews, *Genesis 1–11:26*, 276.
64. Smith, "Agency and Adaptation," 107.
65. Krebs, "Morality," 151.

manipulate weaker and more selfless members into giving more than they take, creating an excess. Selfish members take advantage of this excess by taking more than they give. If however, all members of the group are equally selfish, the group will be embroiled in constant fighting over the division of available resources.

Dominant/subordinate relationships evolved over time to prevent these conflicts, along with clear rules to regulate these relationships, allowing the group to survive over longer periods of time. Once the group is solidified through these rules, they become sexually intertwined, and Darwin's principle of altruism as a way of preserving "close kin" genetic material takes over to produce altruism. Moral codes are then developed to control the intrusion of selfishness that can potentially break down the group, undoing its benefits to survival.

In Krebs's view, then, relationships form for purely functional reasons. Social structure forms to preserve the group, and morality forms to prevent the social structure from becoming abusive, which would lead to the destruction of the group and the loss of its functional benefits. Finally, altruism evolves because the group becomes sexually entangled, and group members act for the functional advantage of preserving their genetic code by protecting the group. According to Krebs, "Moral beliefs and standards are products of automatic and controlled information-processing and decision-making mechanisms,"[66] and these moral beliefs serve strictly functional ends.

Other examples include Joseph Carroll, who holds that altruism has never evolved. Instead, cooperation evolves to support survival, and the rules required for effective cooperation are eventually internalized through each individual's association with the group, eventually becoming moral standards.[67] The purely functional nature of relationships is also found in economic theory, as well, where property rights are tied to the right of first possession,[68] and the person is treated as a "stimulus-response machine," whose choices are "merely the result of a mathematical calculation based upon the environmental constraints imposed."[69]

The most striking difference between naturalistic and Christian anthropologies is in the area of relationships. In the Christian view, even the most selfish or self-centered actions are turned to the service of others. For

66. Krebs, "Morality," 149.

67. Carroll, "Evolutionary Social Theory."

68. Eswaran and Neary, "An Economic Theory," 203.

69. Cleveland, "Connections Between the Austrian School of Economics and Christian Faith," 664.

instance, even the act of eating, which is a matter of personal pleasure and sheer survival, is turned into a communal act of giving to others in relationship. Naturalistic anthropologies, however, take every altruistic act and turn it inward, towards self-survival and functional gain. Neurodigital media designed and developed within a naturalistic anthropology, then, will turn all acts, even the most apparently altruistic ones, into functional demands for gain in some way.

The Idea of Progress

A chronometer, or clock, has a round face, just as Chronos, the ancient Greek god, was often depicted with a zodiac wheel. Time, in both cases, goes *around* in a never-ending cycle—the aeon in ancient Greek literature, or eternity. This cyclical view of time was combined with fate, something individual humans cannot control because they are in the hands of fickle gods. Within time, there is also *kairos,* the ideal moment in time to take some action to reach success—the moment to release an arrow to ensure it penetrates the target, or the moment the shuttle should be passed through the loom to create cloth.

For the ancient Greek, there is no concept of sin, or the fallenness of humanity. Each person can progress as fate allows, and as they take advantage of right moments (*kairos*) in their lives to advance in knowledge, power, or happiness (depending on the philosophical school). Regardless of the progress of any given person, civilizations, people, and ideas pass through a cyclical pattern in an eternal frame, much like the rising and setting of the sun within a day, the seasons within a year, and the years within the aeon.

The Jewish view of time is far different. There are still cycles of days and seasons, marked off by festivals and worship. Israel had its weekly Sabbath, a time of worship and communion with Yahweh. There were three spring feasts celebrating God's work of freedom (Passover) and the two spring harvests. There were also three fall feasts celebrating God's work of freedom (Sukkot and Trumpets) and atonement (Yom Kippur). In the middle of these six feasts is the Feast of Weeks. Beyond these yearly feasts, Israel had longer cycles, such as the Jubilee Year, in which all debts were released, and land returned to the family from which it came.

These cycles are not, however, the final Hebrew view of time. Each of the seven feasts point to the future as well as the past and present. Passover foretells a Messiah who will do more than rescue Israel from Egypt, Trumpets speaks of God's future victory as well as those in the past, and Yom Kippur speaks of a deeper and more permanent spiritual salvation, rather than a primarily physical one. Time, in the Jewish view, goes beyond the cyclical—there is an ultimately purpose in God's creation, a time when the world will "end"—the cycles will continue, but in a fresh place, a "recreation" of sorts.

Christian theology picks up this theme of an end-of-times more explicitly, calling for the return of a Messiah to rule over a kingdom that will last forever. J. B. Bury says Augustine epitomized the Christian view of time in *The City of God:*

> In Augustine's system the Christian era introduced the last period of history, the old age of humanity, which would endure only so long as to enable the Deity to gather in the predestined number of saved people. . . . The medieval doctrine apprehends history not as a natural development but as a series of events ordered by divine intervention and revelations. If humanity had been left to go its own way it would have drifted to a highly undesirable port, and all men would have incurred the fate of everlasting misery from which supernatural interference rescued the minority. . . . [T]he doctrine of Providence, as it was developed in Augustine's *City of God*, controlled the thought of the Middle Ages.[70]

While cycles within time will continue, there is a definite end, or even goal, to the process of history.

This goal is redemption, or perhaps even salvation. For in the Judeo-Christian view of time, man is also fallen, a sinful creature that has fallen from perfection and must be restored. In Christianity this is the doctrine of original sin, which says that since "every child will be born naturally evil and worthy of punishment, a moral advance of humanity to perfection is plainly impossible."[71]

Combining the individual progress of the Greeks with the Judeo-Christian end goal of creation creates the idea of progress: *the world can be constantly improved, creating a more perfect world (reaching towards the Christian ideal of redemption) through progress in individual lives.* As Bury says, "The idea of human Progress then is a theory which involves a synthesis of the past and a prophecy of the future. It is based on an interpretation of history which regards men as slowly advancing—pedetemtim progredientes—in a definite and desirable direction, and infers that this progress will continue indefinitely."[72] For progress to be *progress,* it must not be under the

---

70. Bury, *The Idea of Progress*, 21.
71. Bury, *The Idea of Progress*, 21.
72. Bury, *The Idea of Progress*, 3.

control of some external force, or providence, but rather under the direct control of humanity—humans must, in essence, lay hold of the engine of evolution and shape it towards a desirable end.

Progress and the Person

To reshape society towards the end of progress necessarily requires two things: a view of the person that allows some humans to reshape others, and a set of tools or techniques to perform the operation. The first of these, a view of the person allowing (or perhaps even demanding) the reshaping of individuals towards a better society, is supplied by a naturalistic anthropology. While Christian anthropology holds that dignity, freedom, and relationships are grounded on something outside the person, and therefore inviolable, naturalistic anthropology holds that all three arise from within the person or culture.

All technologies—including neurodigital media—developed using a naturalistic anthropology will hold human dignity is situational rather than absolute. Violating a user's dignity in order to achieve a noble goal at a social or personal level will be considered acceptable. For instance, violating a user's dignity to discourage a habit (such as smoking) is justifiable so long as the user is perceived to be better off within a naturalistic anthropology.

In the same way, naturalistically grounded neurodigital media will consider moral freedom illusory or fundamentally compromised, and hence will find it easy to justify truncating moral freedom if some "greater purpose" is in view. For instance, an advertiser might consider it morally justifiable to use techniques that direct users along fast thinking to achieve a sale, so long as they consider their product helpful—or at least not directly harmful.

The second requirement is a set of tools and techniques that can "make possible human mastery over time and place, and correlatively over nature and human nature,"[73] an objective that ignores C. S. Lewis's prescient warning, "what we call Man's power over Nature turns out to be a power exercised by some men over other men with Nature as its instrument."[74] The next chapter explores the technologies that enable just this sort of control—the explosion of digital computing.

73. Waters, *Christian Moral Theology*, 18.
74. Lewis, *The Abolition of Man*, 55.

4

Creating the Digital World

NEURODIGITAL MEDIA MUST NOT only have a vision of reality, it must also have a physical form—an embodiment in the real world so designers can create it, developers can build it, and users can live in it. As Andy Crouch points out, "Only artifacts that leave their inventors' studios and imagination can move the horizons of possibility and impossibility."[1] While users tend to think of services like Facebook and LinkedIn as being purely digital affairs, these services are software that runs on physical computing, storage, and networking systems, and have physical addresses. You cannot "see" a social media network any more than you can "see" a personality, but you can see the servers and networks the social media network runs on in the same way you can see a person's body.

This chapter traces the development of machine computing (or computing machines), moving back in time to trace their development in parallel with the rise of the naturalistic vision of reality. Most histories of computing consider their three primary components: the ability to calculate using numbers, the ability to store and retrieve information, and the automatic execution of a sequence of operations. Roughly speaking, these correlate to modern processors, storage (such as hard drives), and software. The fourth invention of great importance to modern computing is the high-speed transfer of information between computers or networking.

Behind this physical hardware is the shape of information. How to understand the world is a branch of epistemology or the study of how we know

---

1. Crouch, *Culture Making*, 39.

what we know. For digital computing to succeed, the view of *everything* needed to shift from being framed by narrative to being framed by "facts."

Calculating Machines

Even before the invention of the electrical computer, various aids in performing complex calculations were widely used. For instance, even in the ancient world, pebbles or *calculi* were used as an aid to adding large numbers; the abacus is an example of a portable set of "pebbles" (replaced by beads) placed together in a particular configuration as a mathematical aid. Counting boards, or countertops, were also common. The most challenging problem to solve, however, was carrying a digit across columns in the calculation. For instance, computing 99 + 1 requires the addition of a new column to represent a larger series of numbers.

Wilhelm Schickard invented the first solutions to this carrying problem in 1623, as did Blaise Pascal in 1642. Leibniz, known mainly for his contributions to philosophy, extended Pascal's machine to multiplication as well as addition. Charles Babbage, in the 1830s, sketched out a fascinating Analytical Engine that could perform much more complex mathematical operations. Babbage could never build this device, however, because it was simply not possible to create the mechanical parts required. While Babbage's simpler device, the Difference Engine, has been built in modern times, the Analytical Engine is still out of reach for modern material science. All these machines, however, required the human operator to set the problem. Setting required breaking the problem up into discrete steps the mechanical device could accomplish and carrying the results of each subset of the problem to the next stage. There was no way to store information in one stage of calculation and retrieve it for use in a later stage, nor was there any way to provide a set of instructions the machine could follow across time.

Early Computing: Before Silicon Valley

These inventions did not take any commercial form until the 1880s and the successful Felt Comptometer. At around the same time, William S. Burroughs invented a commercially successful adding machine in the United States, and Brunsviga and Odhner sold such adding machines throughout Europe.

The solution to storing data across calculation stages and storing instructions the machine should execute to solve a problem was solved by Joseph-Marie Jacquard, who created a loom with punch card control.

This loom could weave patterns unachievable by all but the most skilled operators, starting a revolution in machine-controlled fabric production. When building the automation systems used to enable the 1890 census in the United States, Herman Hollerith developed a machine that could perform multiple stages of arithmetic using the same concept of punch cards (though it is not known if Hollerith picked the idea up from the Jacquard looms or invented the idea separately).

The United States Census was not the only organization that saw value in these new computing machines. In the 1830s and 1840s, companies formed to sell a new kind of insurance—life insurance. During the Industrial Revolution, many families moved off farms and into cities, where the men started working at factories and other jobs. In this process, the father became the primary source of income—a family without a father could quickly find itself in dire financial straits. According to Dan Bouk, life insurance companies could sell such insurance by "turn[ing] a life into money."[2] To be profitable, companies could not pay out more in claims than they received in insurance payments, so they sought ways to determine if a person was an acceptable risk—was the person financially stable (able to pay the premiums indicated), physically healthy, and not likely to commit suicide (they would pay premiums long enough to cover any eventual payout on the part of the insurance company).

The companies selling these policies needed some way to assess correctly, or "make," the risk they were accepting. Getting the risk wrong would either mean a failing business or "too much profit," bringing the watchful eye of regulators. Correctly assessing risk required large amounts of mathematical computing power, so the life insurance industry quickly became a large consumer of computing machines, driving their development and adoption.

From Teletype to Digital Computing

Edward E. Kleinschmidt combined the keyboard and "printer" of an ordinary typewriter with the ability to send data through a wire to develop the teletype. AT&T purchased his company in 1930, and Bell Labs took up improving this technology for broader use. Combining telephone switching technology with the teletype led to a wide variety of high-power computing and communication devices. The teletype survived through the 1970s when the machines were adapted to provide the user interfaces for the first silicon-based computers, both as terminals for larger computers (called

2. Bouk, *How Our Days Became Numbered*, preface.

mainframes, after large telephone switches that were large "frames" of physical switches and wiring) and smaller computers (called minicomputers, for shared use, and microcomputers, for individual use). The teletype lives on in culture today through the "wire services," which first shared news through the teletype system (or across the "wire"), the sound of the teletype machine associated with newscasts, and the short symbols used for trading and reporting financial instruments (such as stocks and bonds).

Earlier computers often tried to compute problems using base 10, or using the digits 0–9, and carrying over columns much as a human might. In this period, however, the idea of computing using base 2, or just the two digits 0 and 1, arose. Because these systems only use two digits, they are called binary. It might seem counterintuitive to move from what seems to be a more compact format humans can directly understand to a format that consumes much more space and that humans cannot understand—but moving from base 10 to binary made the creation of digital calculating machines much simpler. The figure below, will be used to explain.

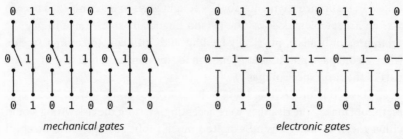

mechanical gates   *electronic gates*

To perform a math operation, a number is input at the top of the figure, where a line with no power applied represents "0" and a line with power applied represents "1." The mechanical switches are then set to one of two possible positions, with "0" represented by the switch being open (or off), and "1" represented by the switch being closed (or on). The result can be read at the output stage—any line where power is present is a "1," and any line where power is not present is a "0." This simple circuit represents an eight-digit, or bit, Boolean AND. By breaking complex mathematical problems into steps based on one of the various Boolean operators, and stringing those steps together, the computer can quickly calculate a math problem of any complexity.

World War II showed the vast computing machinery available was deficient in two respects. First, these early computing machines relied heavily on computers, the people (usually women) who ran the machines to follow a plan laid out on paper to complete the calculations. While these machines could perform the calculations, they could not be programmed in the modern sense; once the result of a single calculation was known, a person (the computer) had to move the result to one of several possible next steps. The computing machine could not move from one step in the calculation process to another on its own. Because the computing machine could not move information from one step to the next, vast numbers of people were involved in moving the results of one stage of calculation to another—people who could be used much more effectively in a time of war. Second, the process of moving calculations along in this way required a lot of time—time no one had when figuring out where to move a ship, manage a fire control system, or decide how many tanks or rifles to build.

Even a fully digital computer using binary math, such as one described by Charles Babbage, could not solve the problem of moving between calculation steps. In 1936, Alan Turing conceived of a digital computing machine with two kinds of memory. The first would be used to store the data being operated on (or being used to calculate results). This memory was similar to the paper punch cards, the ticker tape, and the information transmitted over the wire in a teletype system. The second, however, was a set of instructions, much like the instruction sets given to the human computers who moved information between stages in a calculation. Turing called this combined machine a computer, which is probably the first time it was used about a machine rather than a person. Turing went on to use these principles to build a machine that could decode messages the German army sent through the famous Enigma encryption device, playing a pivotal role in the Allied victory and the shape of history since.

Silicon Valley and the Rise of Electronic Computing

In the mid-1940s, the mechanical switches were replaced by vacuum tubes, which could perform the same operations more quickly and with less power, as shown on the right side of figure 1 above. When using vacuum tubes to build these gates, the mechanical switch in the middle is replaced with a grid; if the grid has power applied, the power from the input cannot "jump" to the output. If the grid does not have power applied, the power from the input can "jump" to the output. The number of switches in parallel is called the bit depth of the system; if the system can calculate across eight binary digits (bits), it is an eight-bit system. If a system can calculate across numbers sixty-four bits in length, it is a sixty-four-bit system. Sixty-four-bit systems can find the results of a problem four times faster than eight-bit systems simply because they can calculate across larger numbers, reducing the number of stages needed to solve a problem.

The ENIAC, unveiled in 1946, was the first digital computer built using electronic, rather than mechanical or electromechanical components. Employing 18,000 vacuum tubes, it could calculate the trajectory of a projectile (fired from a cannon or other large weapon) faster than the projectile could reach the target. The ENIAC was not truly a computer in the modern sense; to change the problem it could solve, the device had to be rewired. This rewiring was similar to the job humans once performed by manually carrying the results of each stage of a calculation to the next stage. Things were getting faster, but still not fast enough to create an entirely digital world.

The UNIVAC, the Universal Automatic Computer, followed the ENIAC in the early 1950s. The UNIVAC had both the ability to perform digital calculations as described above *and* to read a set of instructions out of a second set of memory, performing them without needing to be rewired. With 8,000 vacuum tubes, the UNIVAC heralded the beginning of the kinds of computing power needed to build an entire virtual world—but things needed to be faster, smaller, and less expensive still.

In 1947, Bell Labs invented a device that could perform the vacuum tube's job in digital switches (and hence computers)—but it was built entirely out of silicon. They dubbed this device the transistor. Over the next few decades, advances in transistor manufacturing led to the development of integrated circuits, which placed thousands—now tens of millions—of transistors on a single chip. The invention of the transistor ultimately enabled the development of personal computers, finally leading to a world where the hardware required to build entire virtual worlds and neurodigital media was widely available.

Networks and the Internet

The need to connect computers was evident from the very beginning—the ability to transfer calculation results from one computer to another would allow a life insurance agent in one city to learn about the risk factor of a person they were considering covering from some other location, such as a central office. Connecting terminals, both locally and remotely, allowed the power of large computers—when computers were expensive to build and maintain—with a lot of users. Initially, however, each computer system had its own networking system. If you purchased a Burroughs computer, you purchased Burroughs terminals to connect to it, and it could only talk to other Burroughs computers. The same was true of IBM systems, or any other commercially available system.

Beginning in the late 1960s, however, the US government began connecting research centers in a few coastal cities together into ARPANET. The ability to share computing resources, data, and ideas more readily than other forms of communication quickly became apparent, so the ARPANET was expanded to more locations, and the US military founded the MILNET to connect military research centers. These connections were generally made by purchasing connectivity from large telephone providers—leased lines.

In the early 1970s, Vint Cert and Bob Khan were working on connecting several computers directly over a satellite link, and then through a wireless network carried on a van. They quickly discovered connecting computers over multiple kinds of physical links—wireless in one case, a satellite link in the other—was nearly impossible. Because of this, they began working on a set of protocols that could interconnect multiple kinds of computers over multiple kinds of physical connections—the Transmission Control Program. This ability to interconnect multiple kinds of computers led to a flurry of activity in the research community, with problems being discovered and solved quickly.

For instance, it was quickly discovered that being able to send long text messages between researchers would be useful—so Dave Crocker and a group of others developed and deployed email. Computers needed an address to describe where they were attached to the network, so Internet Protocol (IP) addresses were invented and standardized. People cannot remember IP addresses very well, and computers can move around on the Internet, so a naming system was invented and deployed—the Domain Name System (DNS)—by Paul Mockapetris and Paul Vixie.

Commercialization of the Internet, and its eventual widespread use, was driven by large-scale online services such as Prodigy in 1984 and America Online (AOL) in 1985. These services brought email and near real-time communication through chat rooms, photo sharing, and other services to users connected via MODEMs over Plain Old Telephone Service (POTS) lines, showing that these kinds of businesses could be commercially successful. In 1986, the National Science Foundation in the United States made several critical decisions, such as standardizing on the TCP/IP protocol suite for interconnecting computers and gave MCI, a large telephone provider, funding to create a high-speed general-purpose network that could be used to interconnect regional networks outside the research community. This network, combined with the commercial success of companies like AOL and Prodigy, inspired companies to find ways to provide new kinds of service they could sell based on basic connectivity. As more companies connected to this new network of networks—the Internet—the value of connecting increased. The rapid growth of the Internet from the early 1990s onward is an early example of the network effect creating a system where the value is greater than the sum of its parts.

The power of computing has grown with the number of transistors and other components per Moore's Law, doubling each year from 1965 to 1975, and doubling every two years after 1975. While the power of a computer is only roughly correlated to the number of transistors (and other components) that can be packed on a single chip, moving from the 15,000 gates (where each gate was a vacuum tube) in the ENIAC to several million gates on a single chip gives some sense of the increase in computing power. Individual organizations now operate networks to which millions of computers are attached, where each computer can have twelve or more "cores," each core having millions of transistors and other components.

The Shape of Data

By the 1980s, the world had increasingly large amounts of processing power, and the Internet was increasing in size. If all this capability were just focused on performing advanced mathematical calculations, displaying some pretty graphics, and providing some communication services (such as email and chat rooms), the growth of the Internet would probably have leveled off in a few years. The foundation for the virtual world was built, but for neurodigital media to take center stage, a complete change in the way truth was understood was necessary.

The Binary Tree

The seeds of this radical change can be found long before the computing revolution, within the story of creation. In Genesis 1, Adam is presented with each of the living things, classifying them into kinds such as beasts of the field, birds, and sea creatures (Genesis 1:26). In these divisions, Adam relies on where each kind of living thing lives, its method of locomotion, and other factors as stand-ins for the entire living being; the being is abstracted to one or more of its parts or actions. This abstraction is a common theme in classification and factification systems of all kinds—classification necessarily omits and simplifies to create an easy-to-understand and logically arranged ordered set.

For instance, books in a library are classified by fiction and nonfiction, then within fiction based on the author's name and the style of fiction, then within nonfiction based on topic and subtopic. These divisions tell readers little about the target audience (academic or popular) or the original language of the work. Some things must be left out to create an effective organization system. Multiple classification systems can be designed, but each system will contain only a subset of the available information about any classified object. The reduction of information to create a coherent set is a form of abstraction, a key component of complex systems, and a critical engineering skill.

Based on these passages in Genesis, Linnaeus classified plants and animals into orders, kinds, and species. Linnaeus classified plants based on their reproductive systems and held that the created orders represented the Aristotelian forms.[3] Darwin used this "tree of life," combined with other observations, to posit the theory of evolution; his *Origins* contains a tree-like chart illustrating his theory:

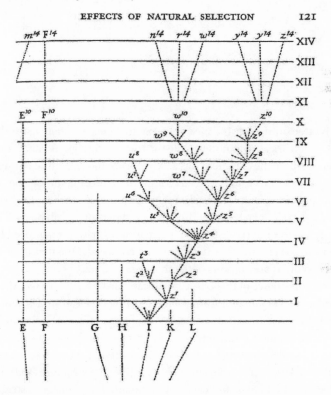

EFFECTS OF NATURAL SELECTION   121

The tree-like structure in Darwin's illustration is clear; he has drawn the root node towards the bottom of the page, with the branches and leaves moving towards the top of the page. Most tree-like structures are drawn with the root node at the top or left side of the diagram, making it more

3. Pearcey and Thaxton, *The Soul of Science*, 102.

difficult to "see" the tree, but the principle is the same no matter where the root node is located.

The tree-like structures are important in the field of computer programming because they can easily be represented through binary numbers. Two forms of trees are illustrated below as they might be represented using binary numbers.

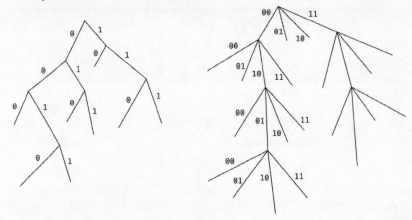

The tree on the left is called a binary tree because there are only two possible branches at each node, while the tree on the right is called a multiway tree. Assume the computer has as a reference to some piece of information the number 0011, and it needs to find this information on the binary tree. The process is simple—the first number is 0, so branch left; the second number is 0, so branch left; etc. The same kind of process can be used in the multiway tree on the right. Given a piece of information referenced by the number 00100111—the first two digits are 00, so take the first branch; the second two digits are 10, so take the third branch, etc.

The importance of this observation can hardly be overstated: the most efficient way to organize information so a computer can find it quickly is to sort it into some form of a tree. All problems of sorting, classification, storing information for fast retrieval, and many others resolve to the single problem of understanding how to sort information into a tree-like structure of some kind. The importance of these tree-like structures to the art of software development is indicated by the many kinds of these structures, and the many sort of algorithms used to create them. For instance, there are radix, Patricia, ordered, unordered, and red-black trees—and all of these formed as trees, as well. There are bubble, linear, random walk, and quick sorting algorithms, which essentially create a tree out of any data set (different algorithms will be more or less efficient on different data sets), and there are specialized algorithms (really heuristics) to find paths through tree-like structures, such as Dijkstra's, the Diffusing Update Algorithm, and Bellman-Ford.

The Social Graph

While these tree-like structures are interesting, they need to be applied to people in some way to create a foundation for neurodigital media. This application follows the path from the social network to the social graph. Social networks form any time people form relationships. For instance, consider a local coffee shop. The owner of the shop has relationships with the owner of the building, a group of people in local government who interact with local businesses, the manager hired to operate the store, at least some of the store's employees (perhaps all of them), people in the various companies who sell supplies to the shop, friends, family, others who attend the same church, and many others. These relationships are the shop owner's social network.

Each of the people the shop owner has a relationship with has relationships with many other people. For instance, the shop's manager may have relationships with the person who delivers supplies from each vendor regularly, and relationships with each of the employees in the shop, and many of the customers who pass through the shop regularly. The manager also has relationships with their family, friends, and many others.

While this social network exists in the pre-neurodigital media world, it is not (usually) explicitly mapped out for any reason—other, perhaps, than by investigators trying to understand a crime, social directors, the older folks in any given family (who know who is related to whom across several generations), and nosy neighbors. If, however, everyone in a community or culture could somehow be convinced to expose their relationships in some way, then these social networks could be exposed and mapped out, as shown below.

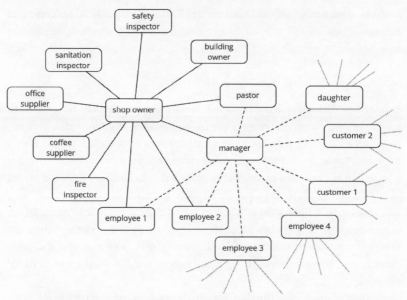

If each person is modeled as a node in one of the tree-like structures described above, their relationships can be modeled as the edges or the connections between the nodes. The result is called a social graph. The shop owner is at the center of an extensive network of people, some of whom are directly connected, such as the office supplier, and others who are second-degree connections, such as the daughter of the manager. Exposing this network of relationships exposes a lot of valuable information, such as who can influence a community or group of people, what beliefs any given person in the network likely has, what ideas or products might be attractive to them, etc. For instance, if you want to influence a person's political beliefs, a good place to start might be with their pastor. How can you influence their pastor? Look through their social network—you are likely to find someone who can.

How do you convince millions of people to expose their real-world relationships in a virtual world? You offer them services for "free," and convince enough people to join your service to cause those who do not to fear "missing out"—both primary strategies of services built using neurodigital media.

Inference and "Intelligence"

It turns out that classification by sorting things into trees can be used to describe much more than social networks and the relationships of various kinds of plants and animals. Language, for instance, can be described in a series of graphs, as anyone who has taken a course where sentences are mapped to understand how the parts of speech relate to one another can attest. Christian scholars even build trees out of theological concepts—systematic theology.

Humans, however, are far too slow at classifying things into trees; the classification process would be more useful if computers could be "taught" to sort just about anything into trees. Computers, however, only know how to resolve questions with yes-no or multiple-choice options. Giving a computer a bunch of pictures and telling it to "mark the pictures containing a cat" seems like something that is not possible. This is, however, precisely the kind of challenge set forth by Alan Turing in a celebrated paper, "Computing Machinery and Intelligence," in 1950.

Turing's Challenge

Turing, in his 1950 paper, set out to explore whether a machine could be considered intelligent. He begins with the definitions of "machine" and "think," saying that "if the meaning of the words 'machine' and 'think' are to be found examining how they are commonly used it is difficult to escape the conclusion . . . is to be sought in a statistical survey."[4] Instead, Turing proposes to replace the question with another one. If a machine takes the place of one participant in a three-person conversation, will the other two participants be able to tell which one has been replaced? If in Turing's definition, a computer can perfectly emulate a person through the course of a conversation, then the machine can be said to be intelligent. Turing calls this the Imitation Game.[5]

One objection raised to Turing's proposal is that the test does not reflect any intelligence on the part of the machine. A counter experiment proposed by John Searle called the Chinese Room problem suggests that Turing has oversimplified the concept of "thinking" to the point where it is no longer useful. In Searle's experiment, a person is placed in a room with a perfect Chinese dictionary. The person is passed pieces of paper with Chinese words written on them, or perhaps phrases, uses the dictionary to translate them, and passes back out pieces of paper containing English (or even some other language). Searle asks if the person in the Chinese Room, who does not know Chinese, is "thinking" in the sense anyone would accept. If the computer in Turing's experiment is doing the same thing—inferring

4. Turing, "Computing Machinery and Intelligence," 433.
5. Turing, "Computing Machinery and Intelligence," 441.

outputs from some sort of "dictionary," regardless of how complex the dictionary might be—it is not thinking.

Mark Halpern raises a second set of objections to the Imitation Game. For instance,

> We do not really judge our fellow humans as thinking beings based on how they answer our questions—we generally accept any human being on sight and without question as a thinking being, just as we distinguish a man from a woman on sight. A conversation may allow us to judge the quality or depth of another's thought, but not whether he is a thinking being at all; his membership in the species Homo sapiens settles that question—or rather, prevents it from even arising.[6]

Halpern further notes that Turing based his defense on a prediction that the meaning of the word *think* would change over time to include computers—but this is not a very compelling argument, either.

Turing's prediction that the word *think* would shift, over time, to include computers illustrates the third objection to Turing's assertion that computers can think—that Turing is implicitly treating humans as complex computing machines, rather than as living, thinking individuals. Turing has assumed a naturalistic, reductionistic view of the person like the one considered in chapter 3. In another paper on whether all numbers are computable, Turing directly compares the computer, a person who uses a set of prescribed steps to complete a math problem, to a computable machine; the person doing the computations can be directly replaced by, and is therefore equivalent to, the computing machine he describes.[7] In another place, Turing says the adult human mind is made up of three things: the initial state of the mind, the education to which the mind has been subjected, and any other experiences. He then goes on to say that while we might not be able to create an adult mind, we might be able to create a child's mind, training it to adulthood over time.[8]

Deep Learning

This idea of starting with a trainable machine, much like a child's brain, and teaching it over time to solve problems, is precisely the mechanism used by

---

6. Halpern, "The Trouble with the Turing Test," 43.
7. Turing, "On Computable Numbers," 251.
8. Turing, "Computing Machinery and Intelligence," 452.

modern "artificial intelligence"—called deep learning throughout this book—to solve problems.

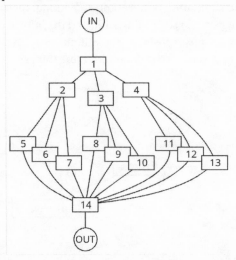

In the above illustration, a series of nodes have been connected to form a network. Assume the purpose of this system is to classify images between cats and other objects. An image of a cat is inserted at *IN,* which injects the image into the network at node 1. Node 1 performs a partial classification and, based on this classification, passes the image to one of the three nodes to which it is attached. Assume the image is now passed to node 2, which then performs some further work, and then to node 5, which then passes the image to node 14, which summarizes the classification work done, providing the output "this is an image of a cat." If the answer is correct, feedback is provided, strengthening the connections between nodes 1, 2, 5, and 14 for this kind of image; if it is incorrect, the feedback provided weakens these connections. The processing taking place at any node in the network does not change; only the weight of the connections between the nodes for any input. Each node in the neural network will perform its job in some slightly different way, such that the network can "learn" which path through the network will correctly classify images (or other data sets).

In a deep learning system, many neural networks are connected, with each neural network becoming a "node" within the system. Each stage of the network, and each neural network connected to the system, can specialize in solving some small part of the overall problem. For instance, in speech recognition, one neural network within the deep learning system may specialize in determining what part of speech a particular word is, or what role that word plays in a sentence. In contrast, another might specialize in understanding punctuation marks and how they alter meaning.

Using Deep Learning on Social Graphs

Now there is the computing power to do something, the structures—trees and deep learning—to do something with, and the information—the social graph created by people exposing their relationships through social media services. What, precisely, can be done with these things?

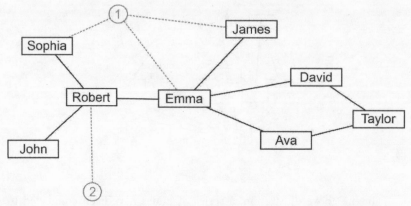

In the illustration above, stronger connections are indicated by heavier lines. One way a system can determine the strength of the connection between two people is by examining their profiles for common interests. For instance, if James and Emma both discuss their love of art museums in their profiles or posts (updates), the system can infer James and Emma have something in common, and probably agree on other things, or share other interests. A second way a stronger connection can be inferred is through shared time and space—if Emma and James often tag one another in pictures, or are often seen together in pictures, they probably have a stronger connection than they would if they do not. A third way is by examining how often James "likes" or "reshares" content posted by Emma. The more often this happens, the more likely it is James and Emma share common beliefs and worldviews.

Given these kinds of information, a deep learning system can predict many things about the people in this simple social graph. Link prediction, for instance, can determine if two people probably know one another, and hence should be connected in the social graph.[9] In the figure above, analysis of this kind might suggest James form a new connection with Sophia because of their shared interest in organization 2—a suggestion made stronger by the relationship between James and Emma, and Emma's connection with organization 2.

9. McPherson, Smith-Lovin, and Cook, "Birds of a Feather"; and Schall, *Social Network-Based Recommender Systems*, loc. 237.

Follow prediction can determine which influence cones a person belongs to—the set of influencers or organizations able to influence the user.[10] In the figure above, analysis of this kind might determine that Robert is influenced by organization 1, and John connects to Robert; therefore, John would probably be interested in following organization 1 as well. Mining social graphs can reveal the size and shape of a community, including those within the community who wield unusual amounts of influence.[11] Information about people and their connections can predict and describe cultural and individual preferences, cultural movements,[12] and even individual decisions.[13]

Conclusion

Chapters 2 and 3, combined with this chapter, provide the background for understanding neurodigital media and systems (such as social media services) built using neurodigital media. Chapter 2 described the intellectual climate in which neurodigital media originated, which is strongly progressive and naturalistic. Progressivism motivates the designers of systems using neurodigital media to impose their view of what culture should look like through these systems. If the designer of a system decides the "world would be a better place" if no one smoked, or everyone exercised three times a week, the progressive worldview motivates designers to build these values into the systems they create.

The second component of the culture of neurodigital media is a strong naturalistic view of the person. If a system designer understands the person—the user of the system—as something that can be changed to fit a particular mold—the progressive mold—they are motivated to find some way to change the user. Naturalism also gives designers and operators permission to impose changes on their users; there is no objective standard of right and wrong, no definition of human dignity standing outside the current cultural context, to consider.

Neurodigital media has one more element not considered to this point—the user experience. This final piece will be defined and discussed in chapter 8 since it is most directly related to truncating human freedom. The next chapter will introduce neurodigital media itself and show how deeply systems built using neurodigital media have penetrated our culture.

10. Schall, *Social Network-Based Recommender Systems*, loc. 251.
11. Grandjean, "A Social Network Analysis of Twitter."
12. Alsaedi, Burnap, and Rana, "Can We Predict a Riot?"
13. Zarsky, "Transparent Predictions."

5

Neurodigital Media

UNDERSTANDING NEURODIGITAL MEDIA—AND THE harm it does to individuals and culture—requires comprehending its culture, its anthropology, and its operation. Chapter 2 considered where the designers of neurodigital media are attempting to take the world—their view of the problems humankind faces, their vision of the future, and their ideas about how to move from the present to their desired future. Chapter 3 covered the results of naturalism on how humans are understood—the naturalistic anthropology. Chapter 4 considered the technologies that allows operators to build virtual worlds based on neurodigital media.

This chapter will describe and define neurodigital media in the context of the background provided by chapters 2, 3, and 4. Neurodigital media is not a term widely used to describe systems and services—after all, this term was only recently coined to describe a set of technologies and a vision of reality—so the second part of this chapter will describe the many uses of neurodigital media. The object of describing these uses is to illustrate how ubiquitous neurodigital media is in modern (or perhaps "technologically advanced") cultures, and hence the importance of understanding and engaging with its impact.

Defining Neurodigital Media

Neurodigital media has three distinct parts. It is *neuro* because it relies on theories of the mind and behavior to shape and mold users' decisions. It is *digital* because it relies on digital technologies and an engineering mind-set

74

to sort, sift, and analyze information about people. It is *media* because it is a form of communication as well as a system. Essentially, neurodigital media is a form of communication that allows the operator of the communication system to shape and mold users' decisions and beliefs using digital technology and techniques.

Neuro

In the 1930s, while a graduate student at Harvard University, B. F. Skinner designed a box—the Skinner Box—that can be used to train pigeons, mice, rats, and other animals. The box normally consists of a speaker, a light, two levers, an electrified floor, and some form of chute or entryway through which pellets of food are given to the animal inside. Skinner theorized there are two forms of behavior in all animals (including humans): respondent behaviors are those actions an animal takes in response to some external stimulus; operant behaviors "come from within" the animal (there is no external stimulus). Skinner placed different animals in the box and then rewarded them for pressing one lever, or "punished" them for pressing the other lever. He theorized that he could elicit operant behaviors through both positive and negative reinforcement, thus proving there is no free will.

The Skinner Box was a complete success. Researchers can train a pigeon to peck at one of the levers, or a rat to press one of the levers, constantly in the hope of receiving a reward or avoiding a shock, a bright light, or a loud noise. One of the critical points discovered through Skinner's research is how often a reward or punishment is critical to conditioning the animal. Producing a reward every time the lever is pressed, called continuous reinforcement, does not seem to cause the operant behavior—the animal will only press the lever when it wants food, rather than constantly. After trying various schedules, Skinner discovered variable reinforcement would consistently produce the behavior he sought. In variable reinforcement, rewards are only ever given in response to the desired action, but they are given randomly. This sort of random reward somehow causes the animal to mentally connect the action with the reward, but the uncertainty of a reward causes the animal to try constantly to receive the reward—so the animal's brain goes into a sort of "zone" where it maximizes rewards by constantly trying the action.

There is no doubt that operant conditioning works on humans. Casinos are large Skinner Boxes, designed to be a complete environment conducive to humans "getting into the zone," focused on receiving the rewards the various machines and games produce. Slot machines (including their online variants) can produce perfect variable rewards. The machine can record the amount of time between each "play," and potentially even use sensors to determine the user's heartbeat, perspiration level, eye movement, etc., to determine excitement level. The machine can also dynamically set "the odds" to control how often the player "wins," perfectly controlling the variable reward rate to manage the emotional state of the player. The slot machine was, until recently, one of the most addictive technologies ever invented.

Multi-user games and social media networks, however, have adopted many of these same techniques to build many more kinds of Skinner Boxes. While users believe their posts, "likes," and "friend requests" are all seen by anyone they are connected to and anyone who follows them, the reality is far different. The system operator can control the number of people who see any given post by filtering it from users' timelines or recording "views" that do not exist. The operator can control "likes" and connection requests in the same way.

Skinner's theories cannot completely explain human behavior, however. People, unlike many animals, can decide to untangle (or break) a habit, for instance. There is clearly more than the results of operant conditioning in response to an environment going on in the human brain. For the naturalist, however, this "something else" cannot be explained through something like the *imago Dei;* there must be a natural explanation. Later theorists, such as Richard Thaler and Cass Sunstein, argue humans have two thinking systems, fast and slow. Operant conditioning works on the fast thinking system, but not the slow one. This leads to the observation that the more often you can cause people to use their fast thinking system, rather than their slow thinking system, the more often you can use operant conditioning to control their beliefs and actions.

By presenting choices carefully—a process called choice architecture—designers can nudge a person toward making a particular choice in any situation. The person who engages their slow thinking process before deciding may choose contrary to the designer's preference, but most people will decide the mental energy required to think slow simply is not worth it. They will just move through whatever process is in front of them, following the bread crumbs carefully laid out by the choice architect.

The topics of choice architecture, nudging users, habit formation, addiction, and user experience will be considered in greater depth in chapter 8.

Digital

Neurodigital media is digital because it relies on digital technology in every phase of its operation. Digital systems are used to collect information about users, including their relationships, their actions, and their beliefs. This information is stored digitally and processed using digital computers to build a "profile" about each user. This profile contains information about what each user believes, what can influence the user, and who can influence the user. This profile can also contain information about the user's optimum reinforcement schedule and their intelligence. While some of this information can be directly obtained from a person's reported income, education, etc., much of it can also be inferred by their grammar, style of writing, how they phrase search queries, and even how quickly they type or how many mistakes they make in typing.

Digital, however, does not just mean how these systems obtain, store, and process information. It also includes a digital engineering, "problem/solution" mind-set. In this mind-set, retrieving information so it can be used in some way is of primary importance. To achieve this end, every piece of information is stored and classified into tree-like structures. Because there is so much information, and humans are not fast at finding patterns in these vast stores, deep learning and other kinds of inference engines, commonly called artificial intelligence, are created, trained, and used. Reality—including people—is reduced to a set of classifiable facts.

In the digital world, everything is binary, either a 0 or a 1. There are no narratives (beyond those expressed in a tree-like structure), no fuzzy middles, and nothing outside or beyond what can be described as a 0 or 1. Since God is neither a 0 or 1, for instance, God simply does not appear in this world. The same can be said of anything human that rises above the physical.

Media

Media, in its simplest definition, is a form of communication. All neurodigital media systems are used to communicate in one way or another. They all have some form of user interface or user experience, they all consume information from users, and they all present information consumed from one user to other users. Communication might be in the form of a post, a like, a connection, or even an alert about a traffic jam, or a product recommendation.

To summarize, the culture of Silicon Valley is progressive, believing human culture should be shaped to create an ideal future. The naturalistic anthropology of Silicon Valley allows this shaping by reducing human freedom and dignity and making all relationships functional. Finally, the tools of Silicon Valley enable the reshaping of individuals, with permission from a naturalistic anthropology, towards the goals of progress. These tools are both mind focused, learning how people think, and technology focused, using digital computing to "shortcut" the thinking process of each user towards a greater goal set by the designer.

Resonance

Neurodigital media, like all other technologies, has a technical and economic structure that results in a content bias. The content bias of each technology reflects constraints on its ability to express reality in its fullness; each medium presents users with a flattened or reduced version of reality. The medium itself mediates the user's perception of reality through its technological constraints. Neil Postman calls this the medium's resonance:

> Every medium of communication, I am claiming, has resonance, for resonance is metaphor writ large. Whatever the original and limited context of its use may have been, a medium has the power to fly far beyond that context into new and unexpected ones. Because of the way it directs us to organize our minds and integrate our experience of the world, it imposes itself on our consciousness and social institutions in myriad forms.[1]

Postman states that television "has made entertainment itself the natural format for the representation of all experience."[2] According to Tim Wu, entertainment was first connected to the selling of audience attention in 1833 by Benjamin Day in the *New York Sun*,[3] and then again using radio through the *Amos 'n' Andy* show on radio and the invention of "prime time" in 1928.[4] By the mid-1950s, television had taken over the role of radio as the entertainment medium of choice, consuming almost the full attention of every person in America during prime time (Wu calls this period "peak attention.")[5]

1. Postman and Postman, *Amusing Ourselves to Death*, 18.
2. Postman and Postman, *Amusing Ourselves to Death*, 87.
3. Wu, *The Attention Merchants*, 12.
4. Wu, *The Attention Merchants*, 87–89.
5. Wu, *The Attention Merchants*, 129.

In the 1970s, a new medium began to rise: the computer network service. While early research networks were restricted to noncommercial use, large-scale bulletin board systems, like America Online (AOL) and Prodigy, were quickly commercialized. According to Mary Meeker, the commercialization of the Internet has resulted in its domination by social media and shopping services.[6] To succeed, these new forms of media needed to cause millions of people to switch their attention from the television to new media services, and then discover how to monetize their users' attention. Wu argues that new forms of media can gain attention in one of two ways: "the first is to present something more compelling than the competition; the other is to slip into some segment of the public's waking life that remains reserved or fallow."[7]

AOL found something both more compelling than television and more able to slip into even the tiniest fragments of a user's time: the private chat room. By 1997, AOL claimed to have more than 19,000 private chat rooms running on its service.[8] Chat rooms, a precursor to the modern social media network, enabled two-way communication between the performer and the watcher. In a world where everyone is a publisher, everyone can be an entertainer. Anyone using the service can gain an audience and the commercial power of selling the attention of their audience.[9] The resonance of these semipublic spaces where anyone can be an entertainer, and each person has an economic or personal interest in gaining followers (an audience), is performance. Jonathan Haidt and Tobias Rose-Stockwell argue that users of social media perform for others because "Social media, with its displays of likes, friends, followers, and retweets, has pulled our sociometers out of our private thoughts and posted them for all to see."[10]

The development of large-scale platforms based on neurodigital media, such as social media, results in what Shanyang Zhao, Sherri Grasmuck, and Jason Martin call nonymous environments. These environments have both online and offline components, and the offline relationship anchors the online persona.[11] They state,

6. Meeker, "Internet Trends 2019"; Huston, "The Death of Transit?"; and White, "Death of Transit."

7. Wu, *The Attention Merchants*, 134.

8. Wu, *The Attention Merchants*, 201.

9. Goldhaber, "Attention Shoppers!"; and Manson, "In the Future, Our Attention Will Be Sold."

10. Haidt and Rose-Stockwell, "The Dark Psychology of Social Networks."

11. Zhao, Grasmuck, and Martin, "Identity Construction on Facebook," 1818.

> [I]n a fully anonymous online world where accountability is lacking, the masks people wear offline are often thrown away and their "true" selves come out of hiding, along with the ta-booed and other suppressed identities. The nonymous online world, however, emerges as a third type of environment where people may tend to express what has been called the "hoped-for possible selves."[12]

This expression of the hoped-for possible self is a type of performance designed to attract positive attention. Some users intend the performance of the hoped-for self to increase their social standing and personal well-being (although it often has the opposite effect, as Mai-Ly N. Steers, Robert E. Wickham, and Linda K. Acitelli have shown).[13] Other users vie to build large audiences[14] and use them for commercial advantage.

When everyone is a performer, there is no longer anyone who pays attention, and attention itself becomes a commodity to be captured and sold. Jason Jercinovic states that the goal is the market of one: the ability to gain and hold the attention of each person, influencing their behavior and beliefs for commercial and social gain.[15] According to Sven Helmer, the market-place of attention may eventually include people paying to avoid screens and advertising;[16] Nellie Brooks maintains human interaction may become a "luxury item."[17] If attention becomes a commodity, then engagement becomes the measure of a platform's value, and seduction (or encouraging addictive behavior) becomes a business model (chapter 4 considers the impact of creating a market in attention on the individual user).

Objectives

In the common parlance of information technology, content and access providers are said to reach eyeballs. Access providers, such as cellphone and cable network operators, are the primary technical contact point between each user and the larger Internet, across which content is carried. Content providers are destinations, or rather places where users to go to interact

12. Zhao, Grasmuck, and Martin, "Identity Construction on Facebook," 1898.

13. Steers, Wickham, and Acitelli, "Seeing Everyone Else's Highlight Reels."

14. These audiences are often called platforms. The use of *platform* to describe both the audience of an influencer and a social media service is confusing, however; so this paper restricts the term to the social media service.

15. Jercinovic, "Markets of One."

16. Helmer, "May I Have Your Attention, Please."

17. Bowles, "Human Contact Is Now a Luxury Good."

with others. Content providers are services such as Facebook, Google, Tik-Tok, and even services like Amazon and Tinder.

Progress

One factor motivating many of the content providers is aligned with the progressive vision of reality—they are trying to make the world a better place. For instance, in June of 2017, Mark Zuckerberg wrote in a Facebook post that, "Right now, I think the most important thing we can do is bring people closer together. It's so important that we're going to change Facebook's whole mission to take this on."[18] He continues, "We have to build a world where everyone has a sense of purpose and community. That's how we'll bring the world closer together. We have to build a world where we care about a person in India or China or Nigeria or Mexico as much as a person here. That's how we'll achieve."[19] LinkedIn, likewise, says, "The mission of LinkedIn is simple: connect the world's professionals to make them more productive and successful."[20] TikTok declares, "Our mission is to inspire creativity and bring joy."[21]

All these mission statements belie a belief in the power of technology to make the world a better place—that if we can all just talk to one another, the world will become a kind of utopia. These mission statements assume people are fundamentally good, as well. Zuckerberg says this directly: "I always believed people are basically good."[22] Many of the people building and operating these services truly hold to the vision of progressive naturalism—humanity is perfectible, and can be perfected by applying the right tools and attitudes in the right way.

Economics

Good intentions alone, however, cannot turn a profit or feed your family. There must be some way to make money through these services if their mission is to succeed. Someone must pay chemical companies to source silicon, semiconductor companies to turn silicon into chips, computer manufacturers to turn chips into computers, networking companies to supply

18. Zuckerberg, "Bringing the World Closer Together."
19. Zuckerberg, "Bringing the World Closer Together."
20. "About LinkedIn."
21. "About TikTok."
22. Zuckerberg, "Bringing the World Closer Together."

bandwidth, and coders to develop the applications necessary to build these virtual worlds. While the early Internet pioneers initially opposed advertising as the primary revenue stream for these services, economic reality soon imposed itself. People would not pay enough to make building the large-scale infrastructures required to make these services possible, and there were not enough volunteers in the world to support the infrastructure. To continue building the systems these pioneers thought would bring a digital utopia, they made a "deal with the devil"—they accepted advertising as their primary revenue stream.

Advertising (discussed in more detail in chapter 7) voraciously consumes attention and time. Like any other "drug," it always takes "more" to get and hold attention over time. As time passes, then, neurodigital media services always turn more strongly in the direction of creating products designed to sell more stuff, moving from the somewhat innocent, like putting advertisements in front of users, to the more difficult to justify, such as dark patterns (considered in greater depth in chapter 8). For advertising to work, the user must be engaged, so most social media services focus on measuring and improving engagement.

Engagement is normally measured in three ways: the amount of time a user spends on the service each day, the number of times the user interacts with posts that appear on their timeline, and the number of times a user "clicks through" an advertisement or other content. Each of these is measured very precisely by the operator.

Of these three, increasing time spent on the service is the most widely measured and associated with success. The more time a user spends on a service, the more time the service is able influence the user, and the more effective the service's influence over the user will be. Spending more time on a service also increases a user's emotional attachment, making it less likely they will quit, and more likely they will become immersed. Users who spend more time on a service also produce more information for the provider— "data exhaust"—which can then be mined for information on how to influence the user and anyone with which they have a relationship.

Increasing engagement is multifaceted. Perhaps the simplest way services keep users engaged is through careful user interface design. The number and placement of controls, the color scheme, the virtual surfaces, sounds, and even haptic feedback (when a device "buzzes" or vibrates) all make a difference in how long a user remains engaged with the service. One less obvious way to keep users engaged is to carry their state from one device to another. For instance, a shopping service might carry search results or items placed in a shopping cart from a mobile device to a desktop device when the user moves from one to the other, or the point in a video where the user shifted from one device to another.

Another less than obvious way to keep users engaged is by making the service operate as quickly as possible. For instance, Reed Hastings claimed in 2017 that sleep is the primary competition for the Netflix streaming video service.[23]

Unintended Consequences of the Quest for Speed

One perhaps unintended—and not well understood—side effect of the content provider's quest for speed is the Internet itself is being subtly reshaped. The illustration below describes the original "shape" of the Internet.

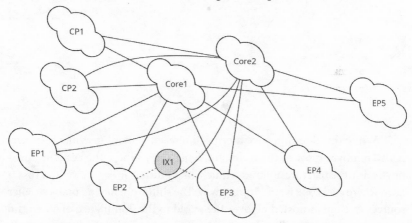

There are four kinds of organizations in this illustration: core providers, who are also called tier one, or transit, providers; edge providers (EP), which accept connections from individuals and businesses, such as a cellular telephone or cable operator; Internet exchanges (IX), which will be explained in the text just below; and content providers (CP). Before large-scale, neurodigital media-based systems became the focal point of the Internet, most edge and content providers connected to core providers because of their reach—by connecting to one or two of the core providers in the Internet, you could reach virtually every user. There were some geographic areas where a group of edge providers would get together and create an Internet exchange, so they could carry "in region" traffic without passing through a core provider. A key point is that core providers charge to carry traffic between other providers (they transit traffic, hence the alternate name transit provider).

23. Raphael, "Sleep Is Our Competition."

Content providers face two problems. First, they would like to stop paying the core providers for carrying their traffic to users. Second, they would like to make their service quicker, so they would like to get their data closer to their users, and reduce the number of networks (or hops) through which they must pass to reach their users. To resolve this, content providers began connecting directly to larger edge providers and Internet exchanges.

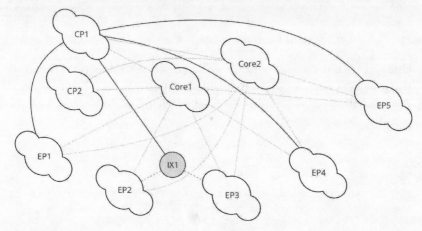

While this does improve the speed of each content provider's service, it also removes most of the traffic from the core. Where the content flows, physical infrastructure follows—much like putting in a sidewalk across a lawn where everyone is walking anyway. The shift from having many smaller sources of content hosted all over the world to the dominance of a handful of content providers is causing the physical reshaping of the Internet itself. One of the many problems with understanding how to undo the effects of large-scale content services built using neurodigital media is deciding how to undo the centralization of the Internet's physical infrastructure.

Pervasive Neurodigital Media

Readers reflecting on these first five chapters describing the history and operation of neurodigital media, as well as the vision of reality it embodies, might end up equating social media, such as Facebook and TikTok, as the primary—or even only—examples of systems built using these technologies. Neurodigital media systems, however, are almost ubiquitous in technologically developed cultures. This section provides some examples of systems that use neurodigital media, some of which may be surprising.

Social Media

Social media, such as Facebook and Twitter, are the most obvious implementations of neurodigital media. These services directly consume information about their users and turn it into direct advertising, social action, and information sold to a wide range of organizations. Most of these companies consider information about their scope and scale confidential, but it is still possible to estimate their size.

Social media services are often measured by the number of users, those who have accounts but use the service infrequently, and active users, those who use the site regularly. In the middle of 2018, LinkedIn reported having 575 million users, 260 million of which use the service at least once a month.[24] By comparison, the United States has a population of around 330 million people—so LinkedIn, one of the smaller social media services, is quickly approaching having more people log on every month than the population of the United States. YouTube has over 2 billion active monthly users.[25] Facebook measures active users daily rather than monthly, with 2.47 billion users logging in to one of Facebook's various social media services at least once a day.[26]

Other ways to measure a social media service is by the amount of time users spend and how much they interact with the service once they have logged in. The average user spends about 11.5 minutes on YouTube each day, and users upload about 500 hours of video to YouTube every minute.[27] According to a recent survey, "34 percent of teens [in 2012] used social media more than once a day; today, 70 percent do. In fact, 38 percent of teens today say they use social media multiple times an hour, including 16 percent who say they use it 'almost constantly.'"[28] Social media services clearly consume a great deal of the combined global attention span each day. A 2019 study showed that social media has largely obliterated book reading among teenagers—presumably in part just because of the amount of time this group spends on social media and playing computer games.[29]

Consuming this much attention leads to large profits—another way to measure the scope and scale of social media services. While it is hard to truly estimate the value of the personal data collected by these companies,

24. Osman, "Mind-Blowing LinkedIn Statistics and Facts (2020)."

25. Cooper, "23 YouTube Statistics That Matter to Marketers in 2020."

26. Noyes, "Top 20 Facebook Statistics."

27. Cooper, "23 YouTube Statistics That Matter to Marketers in 2020."

28. Rideout and Robb, "Social Media, Social Life," 3.

29. Twenge, Martin, and Spitzberg, "Trends in U.S. Adolescents' Media Use," 337.

Elena Botella says, "Our data is valuable to Facebook, above and beyond what the company has previously disclosed. Facebook's average revenue per user in the United States and Canada totaled $132.80 in the past four quarters—seven times more than the $18.70 average revenue per U.S. and Canadian user in 2013."[30] LinkedIn, a professionally oriented social media service, earns 87 percent profit leveraging information it has about its users.[31]

The physical size of social media services can also be used to understand their scale and scope. None of these companies discuss the number of physical computers (servers) they have running, but there are estimates available based on factors like the amount of power their data centers consume, etc. The numbers here, therefore, are estimates based on several different factors. LinkedIn, in 2018, had at least 150,000 servers globally, across three large data centers and a number of smaller ones. Estimates place the number of servers Google operated in late 2017 at around 900,000. Microsoft, in this same time frame, is estimated to have around 1.6 million servers in operation.

Many of the social media companies also own physical infrastructure in the form of optical fiber on land and undersea. Land-based cables are difficult to track, so only undersea cables will be considered here. As of 2020, Facebook owned or was the major user of seven different undersea cables.[32] Most of these cables connect the West Coast of the United States with various Asian countries; a small number connect the United States to Europe. Microsoft, Google, and other social media companies also own cables connecting across oceans in various parts of the world.

Social media services wield vast amounts of economic and social influence. LinkedIn generates about half the traffic business-to-business websites receive, and over 35 million users have received job offers through connections on the platform.[33] One recent study has shown that Twitter users have a disproportionate impact on the content of mainstream or major news outlets, saying, "Our results also indicate that the routinization of Twitter into news production affects news judgment—for journalists who incorporate Twitter into their reporting routines, and those with fewer years of experience, Twitter has become so normalized that tweets were deemed equally newsworthy as headlines appearing to be from the AP wire."[34] Oth-

30. Botella, "Facebook Earns $132.80 From Your Data per Year."

31. Levy, "How LinkedIn Earns."

32. Miller, "This Is What."

33. Osman, "Mind-Blowing LinkedIn Statistics and Facts (2020)."

34. McGregor and Molyneux, "Twitter's Influence on News Judgment," 1.

er authors, such as Jonathan Kay and Emily Jashinsky, provide anecdotal evidence of the influence Twitter has over the way news is reported.[35]

In 2012, Facebook conducted an experiment across 61 million of their users to see how much influence the service had over whether users decided to vote in an upcoming election.[36] Sixty million users were encouraged to vote by informing them of local polling places, telling them who in their social network had already voted (this is called social proof, and is considered in more depth in chapter 7), and allowing them to place an "I voted" "sticker" on their Facebook feed. A further 600,000 users were simply given information about where to vote locally. The researchers took the differential in the number of people who voted between these two groups as showing the influence of relationships formed in social media networks. Some 3 percent more of those encouraged to vote did so. This might seem like a small number, but many elections are decided by far less than 3 percent of the vote—and this was a simple campaign, rather than one designed using all the various social control techniques available.

Social Recommender Systems

You add a book to your shopping cart on an e-commerce site, such as Amazon, and a list of recommended books, videos, songs, and even—perhaps—cleaning products appear on a "suggested products" or "frequently bought together" box some place on the page. In some cases, it might seem like the e-commerce service is "just being nice" in recommending something you might otherwise forget—a memory card or batteries for that camera you are thinking about buying. In others, however, there is no obvious connection between what you are looking for and the suggestions.

Do these suggestions work? Do they cause people to buy more? Yes—as Steve Jobs, the founder of Apple says, "a lot of times, people don't know what they want until you show it to them." In fact, according to Shaban Arora, "Because of how well recommendation engines boost subscriber numbers through engagement and stickiness, facilitating such serendipitous discovery has turned into a high stakes multi-billion-dollar race for the world's biggest digital companies."[37] In 2016, 35 percent of Amazon's sales were driven through recommendations[38]—a number that has probably only

---

35. Kay, "The Tyranny of Twitter"; and Jashinsky, "The Tail of Twitter Is Wagging the Dog of Media."

36. Bond et al., "A 61-Million-Person Experiment."

37. Arora, "Recommendation Engines." Jobs is also quoted here.

38. MacKenzie, Meyer, and Noble, "How Retailers Can Keep Up with Consumers."

grown as Amazon's developers have learned how to incorporate ever more information into their system and their data pool has increased in size.

How do these social recommender systems work? The illustration below, taken from an Amazon patent,[39] will be used to explain part of the process.

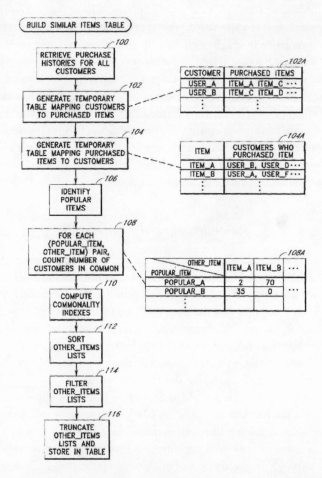

Amazon's recommendation engine builds a list of "similar items" by first searching for what other customers have purchased. If many customers have purchased two items together in the past, this probably means these two items are related in some way. The recommendation engine then searches for the list of customers who have purchased this item. Why should this matter? Because some customers may have more influence than

39. Jacobi, Benson, and Linden, *Personalized Recommendations.*

others, which can be determined based on how many people have found their reviews helpful. Amazon might, at this step, also add information from outside sources, such as social media networks. The information included here might be social characteristics, such as income level, location, hobbies and interests, political and religious beliefs, community membership, etc.

Third, the recommendation engine looks for the most popular items on the list. While "most popular" might imply the sheer number of items sold through Amazon itself, there are many other factors the recommendation engine might also consider at this point. Sentiment analysis is a large and growing area of research for machine learning.[40] In sentiment analysis, review services (such as *Consumer Reports* or *Tom's Hardware*) are mined for articles reviewing items and the comments on those reviews. Further, posts on social media sites are indexed for product names and sentiment. If a product is receiving a lot of positive comments across social media and reviews on influential sites, it is likely "becoming more popular," and hence is more likely to be purchased. Negative reviews or comments will reduce the likelihood of a product being included in the similar items list.

Combining these lists into a single list is where the secret sauce really is. Each factor—related items, social factors, and item popularity—must be weighted. In other parts of the process, the recommender engine must filter out items the user has recently purchased, items deemed inappropriate for the user (if the user is under eighteen, for instance, some products might not be recommended), and then filter based on user reviews (there is no point in recommending products with low ratings, as no one will buy them). A deep learning neural network is used to develop the weights through trial and error; whether an adjacent purchase is made is fed back into the network as training material. Thus, over time, the deep learning network will become more accurate. If each customer's weighted preferences are stored separately, the deep learning network will not only become better in general, but it will develop a good understanding of each customer's preferences at an individual level.

Simply recommending products, however, is not the only way a recommendation engine can maximize profits. Retailers can also use all these factors to determine the highest price they can charge without losing the sale. The customer's sentiment—how much they want this product at this moment in time—and their income level are going to be primary factors in determining the optimal price.[41] Of course, placing products for which

40. Feldman, "Techniques and Applications for Sentiment Analysis."

41. Bhattacharyya, "Pressured by Amazon"; and Kristof, "How Amazon Uses 'Surge Pricing,' Just Like Uber."

Amazon receives a higher profit margin on the recommendation list, instead of items with a lower profit margin, is also important.

Netflix also depends on social recommenders to drive profit, which they describe as a "a collection of different algorithms serving different use cases that come together to create the complete Netflix experience."[42] The Netflix recommendation engine includes the top videos, which videos are trending, and the similarity between videos. Again, Netflix probably does not use information gained only on their site, but likely also uses larger sentiment analysis from across search engines and social media sites to determine which videos to recommend to users. Netflix's recommender engine saves the company some 1 billion dollars in advertising and other costs, allowing them to invest in creating new content.[43]

About 80 percent of the hours streamed at Netflix are influenced by their social recommender system.[44] YouTube has a similar success rate with their recommendation engine, with 70 percent of the content watched on the service determined by its recommendation algorithm.[45]

Other Uses of Neurodigital Media

Social media and social recommender systems are the most obvious uses of neurodigital media, but there are many others. Mapping software—which you probably use to find your way to just about anywhere—is also a form of neurodigital media. Mapping programs consume information about road conditions from users by considering how many of them there are in one place at a particular time, how quickly they are traveling,[46] and where they physically go (which is helpful in finding new roads, or discovering roads are closed). This information is also used, however, to determine the most popular stops along a route, such as coffee shops, shopping destinations, or restaurants. Over time, with enough information, each user's experience can be finely tuned, steering them to local businesses "along the way," and maximizing profit.

Political and social campaigns also make heavy use of neurodigital media, both as a user of existing services and in creating new ones. Joe Trippi argues, for instance, that Barack Obama's presidential campaign "restored

42. Gomez-Uribe and Hunt, "The Netflix Recommender System," 2.

43. Arora, "Recommendation Engines."

44. Gomez-Uribe and Hunt, "The Netflix Recommender System."

45. Cooper, "23 YouTube Statistics That Matter to Marketers in 2020."

46. For a somewhat humorous illustration of the kinds of information Google Maps collects and how it is used, see Weckert, "Google Maps Hacks."

the primacy of the individual voter" by innovating in their use of social media and custom systems—based in neurodigital media—to measure and influence voters.[47] Progressives, however, did not celebrate the use of social and neurodigital media by Donald Trump in the 2016 election—rather they called the efforts unethical and manipulative.

Many companies use forms of neurodigital media in the workplace to make employees more productive, perhaps even reaching the level of encouraging addictive behavior toward their tools.[48] Other companies use information gathered about employees to determine who is likely to quit in the near future, using this information to either encourage them to "move on," or working with them to prevent them from leaving.[49] Biometrics and health data are often collected by companies to improve health outcomes (since the company generally pays for the employee's insurance), to track employee productivity over time,[50] and to improve employee actions for health and ethics.[51]

Reading applications gather information as users read, feeding this information back into social recommender systems.[52] Companies are working on connected homes and cars through the Internet of Things (IoT) that will feed information into neurodigital systems, and present screens, interactive voice systems, and other interaction points between the inhabitant and the system, creating opportunities to build an immersive neurodigital environment.[53] Sharing services, such as Uber and Airbnb, use their rating systems, along with other information collected through their applications, to improve services and build marketing profiles of their users.[54]

Governments are looking at using neurodigital media, as well, for many different purposes. Forms of neurodigital systems are used in predictive policing, in setting bail, setting prison sentences, and in deciding on parole.[55] Cities have created projects to build massive sensor networks, tracking the movement of people, crime, interactions, electricity use, and

47. Trippi, "Technology Has Given Politics Back Its Soul," 34.

48. Wadhwa, "Workplace Technology."

49. Nair, "A Predictive Analytics Model."

50. Mateescu and Nguyen, "Explainer: Workplace Monitoring & Surveillance"; and Mateescu and Nguyen, "Explainer: Algorithmic Management in the Workplace."

51. Fort, Raymond, and Shackelford, "The Angel on Your Shoulder."

52. Lynch, "The Rise of Reading Analytics"; Wicker and Ghosh, "Reading in the Panopticon."

53. Tajitsu, "Toyota, Panasonic to Set up Company for 'Connected' Homes."

54. Paul, "Rating Everything."

55. Richardson, Schultz, and Crawford, "Dirty Data, Bad Predictions."

many other factors.[56] Cities intend to use this information to improve the financial performance of core shopping areas, control traffic flow to relieve congestion, prevent riots,[57] and reduce crime.[58]

Conclusion

Neurodigital media is *neuro* because it relies on psychological insights about the person, including how to structure choices and the user experience to guide users to a specific decision. It is *digital* because it relies on digital computing, binary representations of data, and methods developed using systems ecology for analytics. And it is *media* because it interacts with users through text, images, video, etc. Current systems built on neurodigital media assume a naturalistic view of the person. Finally, neurodigital media is almost ubiquitous, reaching into every corner of our personal lives, including dating, using a commercial ride-sharing service, interacting with others online, and even in interacting with the government.

This is the final chapter introducing neurodigital media, explaining its cultural roots, how it works, and its scope. The next chapter will dive deeper into the clash between the Judeo-Christian and naturalistic views of the person in the area of human dignity. This is where things are going to truly get scary.

56. Jaffe, "Toronto Tomorrow"; and Kofman, "Google's Sidewalk Labs."
57. Enderle, "In the Shadow of Paris."
58. Zhihui, "Nowhere to Hide."

6

Dignity

Genesis 4:1 is among the oddest passages in the Scriptures: "Now Adam knew Eve his wife, and she conceived and bore Cain, saying, 'I have gotten a man with the help of the Lord.'" Whatever can it mean for a person to "get" something with the "help of the Lord"? One way to read this is that Eve prayed to God, asking for his help in having a son, and God answered her prayer. While the sense of the passage may support this reading, the phrasing "gotten *me* a man" seems to imply something more. There is something selfish in the way *me* is used, something that reaches beyond the simple "I have a request, please help me fulfill it" mind-set.

What Eve seems to be saying is that she has *used* God to get something she wanted. It's like this—assume you are back in your childhood, you have a brother named Johnny, and you're told to go out and shovel the snow off the driveway with Johnny. There are at least three ways you can understand this statement. The first is that you are to go out and work with Johnny to clear the snow off the driveway. The second is you are to go out and tell Johnny how to clear the snow—to supervise, or to control. This second sense gets close to a third sense—you use Johnny as a tool to clear the snow.

You might be forgiven if you imagine Johnny, stiff, being used as a shovel to clear the snow. I suppose the feet of a person would make a better scoop than the head, so at least he would be spared the indignity of having his head pushed under the snow every time you want to shovel another load off the driveway. But this gets to the root of the problem—treating a person as a means to an end reduces their dignity. As Kant says, "Man and reasonable creatures exist as a purpose in themselves."[1]

Likewise, as noted earlier, Lewis and Demarest argue humans are always to be treated as an end in themselves rather than as a means to an end: "Persons as spiritual beings are not things to be folded, mutilated, or spindled. As self-transcendent spirits humans are self-conscious and self-determining subjects and moral agents. Christians see in all other persons active beings who should be free from coercion to think, feel, will, and relate."[2] The record of the Scriptures regarding using persons as an end is summarized by Michael Novak, who says, "Christianity made it a matter of self-condemnation to use another human as a means to an end. Each human being is to be shown the dignity due to God because each is loved by God as a friend. Each has God as 'a father.'"[3]

To treat someone as an end in themselves can be expressed in a more helpful way by saying each person should be treated as a whole person. To do otherwise—to treat a person as a tool, or a fount of information, or a source of income—is to flatten the person into something less than a whole person. Doctors flatten human beings by treating them as only a body, and politicians flatten communities by treating them as merely a set of votes to be gained. In many cases, flattening is necessary and useful; the problem arises when human beings are treated solely as flattened representations, rather than as individuals. This chapter will explore what flattening the person means, then considers how users self-flatten, how users flatten one another, and how operators flatten users at scale.

One of the challenges in considering neurodigital media is separating the different kinds of consumers, or answering the question "who is the customer?" In one sense, the individual user is the customer because individuals use the services offered by the operator. For instance, individuals create Facebook and LinkedIn accounts, or use Google Maps to find their way from one place to another. In another sense, the company that builds and operates these services is also a customer of the individual users, collecting data that is then used to create economic value. Advertisers and consumers of the data these operators collect. Here users are individuals and organizations that connect with others through the system; operators are individuals or organizations that collect and analyze information through the system to achieve financial and social goals.

1. Author's translation of "Nun sage ich : der Mensch und überhaupt jedes vernünftige Wiesen, existirt als Zweck an sich selbst." Kant and Kirchmann, *Grundlegung zur metaphysik der sitten*, 52.

2. Lewis and Demarest, *Integrative Theology*, 172.

3. Novak, "The Judeo-Christian Foundation of Human Dignity," 109.

Flattening

The concept of flattening means to take something that has multiple dimensions, such as a box, and break it down into a single dimension—like folding the box flat for storage or recycling. In the same way, flattening a person is to remove dimensions from the person in some way, to see only a part of the individual. Since it is hard to imagine "folding" a person to reduce them from a three- to a one-dimensional object, a more useful term is abstraction.

Flattening is the abstraction of persons into mere objects of study, treating persons as merely a source of instrumental value, and conforming the person to the machine, tool, or system built using neurodigital media.[4] Two examples of flattening often used in neurodigital media is the classification of users and treating the user as a simple information processing machine when creating the User Experience (UX).

Flattening the Person through Classification

Neurodigital media relies on replicating social networks in tree-like data structures through information processing to capture user preferences, predict user decisions, and capture cultural movements, as described in the first chapter. This kind of classification requires reducing a person to a small set of attributes, a form of flattening that impacts the way users interact with one another through neurodigital media as well as how operators view users.

An example of this flattening effect of classification will be helpful. Linnaeus's classification system is a tree-like structure, arranging creatures according to well-defined characteristics. For instance, a translated and "improved" version of Linnaeus's *Animal Kingdom* notes it only considers the "varieties of the mammalia, or animals which give suck to their young."[5] One of the many introductions included in this edition divides animals into natural classes such as: "A heart with two auricles and two ventricles; Warm and red blood," with two subclasses, viviparous being mammalia and oviparous being birds.[6] This classification process picks out one feature at a time, abstracting each creature to that one feature. Considering multiple features requires starting at the root, which uses broad classifications, and moving towards the leaves, containing narrower classifications.

---

4. A term borrowed from Sherry Turkle, *Life on the Screen*, 330. Turkle cites Fredric Jameson as the originator of this term.

5. Linné et al., *The Animal Kingdom*, frontispiece.

6. Linné et al., 29.

Regardless of how comprehensive the tree-like classification system is, such a graph can only contain some information about individuals. For instance, in the case of a person, a graph built using Linnaean principles might note a person is among warm-blooded creatures who give suck to their young, so they are mammals. Moving towards the leaf representing humans, the chart might note persons reproduce sexually, give birth to live young, have hair (rather than scales or feathers), are bipedal, have forward-facing eyes, etc.

Applying Linnaean principles to relationships and people requires reducing them in the same way. Daniel Schall, in his work on social recommender systems, describes several methods used to determine if two users "should" be "friends" (link prediction) based on information contained in a social media network.[7] The first of these considers a triad of users (or objects) within a social graph, with connections as illustrated below.[8]

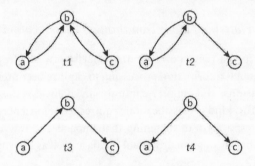

In connection pattern *t1*, *a* and *b*, and *b* and *c* have formed mutual connections. In *t4*, however, *b* has expressed an interest in *a* and *c*, but neither has expressed an interest in *b*. These four cases are arranged in order of likelihood *a* and *c* "should" be connected, with *t1* being the most likely and *t4* being the least likely. Nothing is known about the users represented by the nodes *a*, *b*, or *x* in these four cases other than their connections to other users in the social network. The users are flattened to this one attribute—they are not whole persons with dignity, but objects to be studied and understood to predict and suggest behavior.[9]

While moving from the root of the tree towards the leaves increases what is known about a person, it is impossible for the graph to contain

---

7. Schall, *Social Network-Based Recommender Systems*, loc. 340ff.

8. Adapted from Schall, *Social Network-Based Recommender Systems,* loc. 583.

9. Topirceanu, Udrescu, and Marculescu, "Weighted Betweenness Preferential Attachment"; and Thonet et al., "Users Are Known by the Company They Keep."

the fullness of what being a human is. At the very least, such a graph will not contain moral freedom, dignity, or the relational nature of the person. Persons, when viewed primarily through such a graph, will necessarily be reduced to the set of attributes used for classification in the graph, rather than being treated as a whole person with an essence.

Flattening the Person through the User Experience

While the classification processes used by neurodigital media flatten individual users by reducing them to a subset of a full person, the UX flattens users by treating them as a simple data processing machine. Given some input, the designer should be able to predict the action a user will take; if there is any uncertainty in the action/reaction pair, this means the designer needs more information about the person.

Naturalistic anthropology encourages this form of flattening by removing all intrinsic grounds for dignity and treating the person as the sum of purely physical processes, what Tom Wolfe calls "a piece of machinery, an analog chemical computer, that processes information from the environment."[10] The comparison of the mind to some form of computing machine—the reduction of the person to a machine—is common in the writings of naturalistic thinkers. For instance, Marcin Miłkowski says denying that the mind is just a form of computer is "probably a matter of ideology rather than rational debate."[11]

The equivalence between mind and computer is often implicit in arguments about the nature of intelligence made in the field of information technology. A. M. Turing provides one of the earliest examples in 1936, describing a process where a machine might be created with the intelligence of a child and the ability to learn. Over time, Turing says, this machine intelligence can learn in just the same way as a human child, eventually equaling or surpassing the intelligence of a human adult[12]—implying a mind is a computing machine similar to a digital computer, just instantiated in a different kind of hardware. The equivalence between mind and software, body and hardware, is more explicit in the transhumanist movement, which combines "apparently unrelated movements in biotech, tech, and social justice," to liberate "the human being from the limitations of the body."[13]

10. Wolfe, "Sorry, But Your Soul Just Died," 12.
11. Miłkowski, "Why Think That the Brain Is Not a Computer?," 26.
12. Turing, "Computing Machinery and Intelligence," 60.
13. Emmons, "The Transhumanism Revolution."

Mohammed Bilal explains that users frequently face internal negative emotional triggers; they react to these triggers by "browsing Instagram or watching funny videos online."[14] For Bilal, the user's reactions to negative emotional triggers are involuntary; that these reactions are a result of machine-like states within the mind is implicit. According to Bilal, designers can take advantage of these automatic reactions by redirecting them towards purchasing an item rather than consuming media by capturing a user's attention, causing the user to "give in" because they "anticipate a reward," and thus "engage/invest into the product/app."[15] Bilal describes taking advantage of the negative emotional trigger called the fear of missing out (FOMO). Because individuals are "afraid" of missing some important piece of information or opportunity, they will check email or social media feeds constantly. Bilal created a system that causes users to fear missing out of a "deal" or "sale" on a shopping site, turning the negative emotional trigger into a potential sale for the site.[16]

Bilal describes using the "hook model" to build a flow capturing the user's attention. Nir Eyal unpacks the four steps of this model as trigger, action, variable reward, and investment. According to Eyal, there are two kinds of triggers, external and internal. External triggers are outside the user, in the environment, and contain a call to action in the user's environment, such as an advertisement or an "unread messages" indicator. Internal triggers are generally emotional states over which the designer has no control. The most reliable internal triggers are negative emotional states.[17] Eyal says, "Connecting internal triggers with a product is the brass ring of consumer technology."[18]

To connect an internal trigger to a product, designers must connect the internal trigger to an external trigger, which then promises a (generally emotional) reward if the user follows through on the call to action, such as new content to browse or a product to consider purchasing. The user must take action to receive the reward, physically tying the trigger to the reward. Tying the trigger and the reward together with action creates a physical pathway in the user's mind, taking advantage of neuroplasticity to form a habit. Physical ties of this kind have been reported by Mark Bohay and his colleagues in their research on note-taking, recall, and understanding of

14. Bilal, "How to Design."
15. Bilal, "How to Design."
16. Bilal, "How to Design."
17. Eyal, *Hooked*, 47.
18. Eyal, *Hooked*, 47.

the material.[19] Eyal argues that "doing must be easier than thinking" for habit formation to be successful;[20] the user's rational, or slower, thought processing must be bypassed whenever possible, and the more machine-like process of faster thinking encouraged to be dominant. Eyal says this hook model is used to "connect the user's problem with the company's solution frequently enough to form a habit."[21] This hook model treats the mind as a kind of information processing machine where specific inputs invariably result in known outputs.

Observations on Flattening Techniques

The classification of persons within a tree-like structure—the social graph or network—reduces each individual to those attributes within the classification system, which is less than the wholeness of the person. These classification systems can then treat the person and his or her relationships as merely instrumental means to predict connections, paths of influence, and other information about the user. This trend towards treating the person as a merely instrumental means towards some other ends is continued through the user experience of neurodigital media, which treats the user merely as a means to reaching a goal, such as selling a product or service.

These two techniques, when combined in systems built using neurodigital media, flatten the person to a mere instrument to be used by an operator or other users to reach specific goals. While flattening in narrow circumstances is common, such as a doctor treating a patient as a body rather than a whole person, this treating of the whole person as a mere instrument across all relationships represents a fuller and broader form of flattening. Because the forms of flattening described here involve intimate and personal relationships, they transfer to the way people treat one another outside the immediate context of a service more easily.

Self-Flattening

Just about anywhere you travel, even to your local church, you will find people holding their cameras as far away as possible—maybe even using a "selfie stick"—to take a picture of themselves. A person positioning

---

19. Bohay et al., "Note Taking, Review, Memory, and Comprehension."
20. Eyal, *Hooked*, 61.
21. Eyal, *Hooked*, 13.

themselves for better light, and working to reach that perfect pose and the perfect facial expression—these are all forms of performance.

Taking pictures always contains an element of performance. The parent taking pictures of their child to place in a scrapbook is asking the child to perform. The difference, of course, is that the audience of the selfie is not the family, or the child or parent when they are much older, reminiscing over shared memories.

Selfies

The audience of the selfie is your peers or followers—an immediately accessible global audience. The purpose of the selfie is not to help recover memories, but rather to impress or gain attention in some way. Selfies, then, are uniquely tied to performance.

What impact does taking selfies have on the person? Susruthi Rajanala, Mayra Maymone, and Neelam Vashi note the act of taking selfies is often combined with filters that can modify photographs before they are posted to social media services.[22] According to Samantha Grasso, this means the "pressure to take attractive yet effortless selfies is on."[23] Grasso continues that "studio-quality selfies are as easy as downloading a smartphone selfie app," and then lists eight such apps that allow users to "put their best face forward."[24] One app in the list is AirBrush, which can remove wrinkles, whiten teeth, and enhance "problem areas." Facetune 2 autodetects different areas of the user's face and allows you to "change the curve of your smile, the size of your jaw, the width of your face, resize features, use filters, and more."[25] Using these filters, women tend to slim down their faces, make their face rounder, and increase the size of their eyes. YouCam Makeup allows the user to add a perfect makeup job to their selfie, and Retrica creates an edgy, "old-fashioned" look.[26]

This filtering of the self is a form of what Dean Cocking, Jeroen van den Hoven, and Job Timmermans call "objectification of self by self" by reducing the "self to the construction of appearances."[27] This self-objectification is only possible in the digital world, according to them, because the natural

---

22. Rajanala, Maymone, and Vashi, "Selfies—Living in the Era of Filtered Photographs," 443.

23. Grasso, "The 10 Best Apps for Shooting and Editing Selfies."

24. Grasso, "The 10 Best Apps for Shooting and Editing Selfies."

25. Grasso, "The 10 Best Apps for Shooting and Editing Selfies."

26. Grasso, "The 10 Best Apps for Shooting and Editing Selfies."

27. Cocking, van den Hoven, and Timmermans, "Introduction," 183.

feedback loop provided by real-life relationships can be easily avoided by "pushing a button."[28] Cocking and his coauthors say that in the real world, "the self we construct is met with both resistance and new directions"—elements missing from purely virtual relationships.[29]

Jeremy Rifkin says, "For the more affluent members of society, performance has become far more self-conscious and commercial in nature. Growing numbers of people, especially young people, see themselves as performance artists and their lives as unfinished works of art."[30] Both Nolan Gertz and Rifkin indicate that the resonance of social media—and hence neurodigital media—is performance, a point made here in the chapter 1. Geert Lovink says the slogan "you are what you share" describes the conversion of the person into an "outward facing entity that is constantly reproducing its social capital by exposing value (data) to others."[31]

Gertz describes the act of performing on social media as herd networking, saying,

> The danger of herd networking can be seen in how social networking platforms not only lead brands to act like people but lead people to act like brands, crafting identities and producing content in accordance with the platform-induced need to attain and retain followers, followers whom we cannot be sure are interested in us beyond our content as we ourselves are no longer sure who we are beyond our content.[32]

The self-objectification of the selfie takes place by the user presenting a small part of who they are to the audience instead of "the whole person." Not only is the user presenting less than their entire selves to their audience, they are presenting a modified form of just a small part of themselves—they are flattening themselves by making themselves objects to be admired, derided, or otherwise paid attention to.

Identity Formation

Another way to frame this self-flattening is as a kind of identity creation; selfie-taking is an act that attempts to create a simplified identity representing one aspect of the whole person. One of the most striking results of

28. Cocking, van den Hoven, and Timmermans, "Introduction," 183.
29. Cocking, van den Hoven, and Timmermans, "Introduction," 183.
30. Rifkin, *The Age of Access*, 213.
31. Lovink, *Sad by Design*, 2019, 19.
32. Gertz, *Nihilism and Technology*, loc. 269.

performance is the creation of a narrow identity for each platform. A user may be a social justice warrior on one platform, a consummate professional on another, and a doting mother on a third.

Shanyang Zhao, Sherri Grasmuck, and Jason Martin argue that identity is a self-concept formed when an audience accepts or validates the identity declared or proclaimed by a person.[33] Identity construction in anonymous environments can be separated from the person so that a single person can have multiple identities in different anonymous spaces.[34] However, social media is not a completely anonymous space. Instead, there is some overlap between a user's relationships on each platform and their relationships in real life. Because of this overlap, Zhao and his coauthors call these environments nonymous.[35] In nonymous environments, the user can freely shape their online identity, so that it overlaps with their offline identity, and then validate their identity in both online and offline settings.

In their research, Zhao and company find three forms of identity construction on Facebook. The first is narrative, in which the user primarily discusses themselves. About 37 percent of the Facebook users they investigated had some element of narrative identity construction associated with their profiles. The second identity is the self as consumer, in which the user discusses hobbies and interests. About 73 percent of user's profiles contain these kinds of elements. The third is the self as a social actor, or the self as a performance, consisting of posting images and text where the user is doing something. About 95 percent of the user profiles examined contained elements of the self as performance.[36] The researchers describe this kind of identity as "watch me and know me by my friends."[37]

According to these researchers, users stretch the truth in their self-presentation, creating "hoped-for possible selves" who are "socially desirable entities individuals aspire to be offline but have not been able to embody,"[38] an effect documented by Jamie Guillory and Jeffrey Hancock on the LinkedIn professional social networking service, as well.[39] The projection of these "hoped-for possible selves" is a form of self-flattening. Since the person is presenting only some part of themselves in the most flattering

33. Lovink, *Sad by Design*, loc. 1817.

34. See, for instance, the descriptions of multiple personalities in Turkle, *Life on the Screen*, 267.

35. Zhao, Grasmuck, and Martin, "Identity Construction on Facebook," 1818.

36. These three modes of identity construction are given in a chart at the bottom of Zhao, Grasmuck, and Martin, "Identity Construction on Facebook," 1824.

37. Zhao, Grasmuck, and Martin, "Identity Construction on Facebook," 1825.

38. Zhao, Grasmuck, and Martin, "Identity Construction on Facebook," 1830.

39. Guillory and Hancock, "Effects of Network Connections," 67.

way possible, rather than their whole person, they are self-abstracting. Gaby David and Carolina Cambre find this kind of self-flattening through the construction of a public image in the Tinder dating application, as well, where the production of selfies become "part of a learning process of self-mediation, training users in the editing, curating, and construction of a pseudo-self-authenticity."[40]

Selfie Surgery

Widespread access to filters used to modify selfies, according to Rajanala and his coauthors, has altered "people's perception of beauty worldwide," and expanded the propagation of unrealistic standards of beauty to everyone.[41] According to a study of body dissatisfaction in adolescent girls by Sian McLean and coauthors, "girls who regularly shared self-images on social media, relative to those who did not, reported significantly higher overvaluation of shape and weight, body dissatisfaction, dietary restraint, and internalization of the thin ideal."[42] The researchers found the "strongest correlations in the self-photo sharer subgroup were for photo manipulation and photo investment."[43]

Mills and his colleagues performed similar research, surveying the mood of women who post higher volumes of filtered and unfiltered selfies to a control group who do not. Women who posted high volumes of selfies "reported feeling more anxious, less confident, and less physically attractive afterwards compared to those in the control group."[44] Giulia Fioravanti, Alfonso Prostamo, and Silvia Casale performed a slightly different study to determine the impact of social media use on body image. Their study asked Instagram users to refrain from using the service for a short time to determine if this would impact the subject's self-image; they discovered abstaining from Instagram did have an impact on the body image of women.[45]

Perhaps the most extreme version of self-flattening in response to social media is selfie surgery. Rachel Hosie quotes Dr. Esho of The Esho Clinic as saying, "Previously patients would come into clinics with pictures of celebrities or models they admired and wanted to look like . . . but with the introduction of social platforms and filters over the last five years, more

40. David and Cambre, "Screened Intimacies," 4.
41. Rajanala, Maymone, and Vashi, "Selfies," 443.
42. McLean et al., "Photoshopping the Selfie," 1132.
43. McLean et al., "Photoshopping the Selfie," 1138.
44. Mills et al., "'Selfie' Harm," 86.
45. Fioravanti, Prostamo, and Casale, "Taking a Short Break from Instagram."

and more patients come into clinics with filtered versions of themselves as the goal they want to achieve."[46]

Amy Rose Spiegel and Broadly Staff describe an episode of *Plastic Planet*, a television show that follows "two young women as they go through with cosmetic procedures on their noses in order to feel as comfortable with themselves in real life as they do behind a filter."[47] One of the two young women says the filtering software is "like plastic surgery on your phone."[48] Modifying appearance in the real world to match that seen on a screen breaks down the resistance Cocking and company describe as one of the fundamental differences between the virtual and real worlds; if identity is equally as pliable in the physical world as the virtual one, there is no longer any call to revise our view of our identity.[49] The boundaries of the real world are no longer limits—so long as medical expertise can fill in the gap between the physical and virtual.

Interpersonal Flattening

It is not surprising that users who present a "hoped-for possible self" through social media are then treated by other users as somewhat less than an entire person. The image presented through social media is of a life free of struggle (or struggle only as a performance), rather than life in all its aspects, giving the viewer permission to see falsely a constructed identity as the entire identity.

For instance, according to Steers, "If people portray themselves as happier than they actually are, then perceptions of the happiness and well-being of one's Facebook friends are likely to be distorted."[50] This self-presentation is not, however, the only source of interpersonal flattening in social media services built using neurodigital media. The following sections consider three other sources of flattening: the lack of social cues, the problem of information overload, and the quantification of relationships.

46. Hosie, "People Want to Look."

47. Spiegel and Staff, "I Got Surgery to Look Like My Selfie Filters."

48. Spiegel and Staff, "I Got Surgery to Look Like My Selfie Filters."

49. Cocking, van den Hoven, and Timmermans, "Introduction," 183.

50. Steers, Wickham, and Acitelli, "Seeing Everyone Else's Highlight Reels," 703.

The Lack of Social Cues

Michael-John Turp argues that "social media inhibits the interpersonal emotions normally flowing from reactions to the minds of others," promoting an objective attitude towards others, and making users "less likely to treat others as persons and more likely to treat them as objects or things, which is morally risky."[51] According to Turp, one structural reason for this objectification is the lack of social cues inherent in digital communications caused by the "lack of physical presence," making it challenging to attribute mental states to conversational partners.[52] This lack of social cues is exacerbated by the asynchronous nature of digital communication, removing "powerful elicitors of reactive attitudes that regulate our behavior."[53]

Tinder, a neurodigital media-based dating app, is a quintessential example of the objectification of the other users in a social media service. The Tinder app presents an image of a person, inviting the user to either swipe right to indicate interest, or swipe left to indicate a lack of interest. If two users swipe right on one another's images, an animation plays, and a line of communication is opened between the two users. While the Tinder interface, and its focus on images, have been widely copied, it was unique when it was first released. Tinder's focus on images as the primary means to judge whether to "date" someone else creates what Gaby David and Carolina Cambre call "screened relations of intimacy," which "highlight aspects of speed, ethereality, fragmentation, and volatility."[54] The result is a "mediazation and depersonalization," "encouraged as a result of the speed of profile-viewing enabled by the swipe logic,"[55] allowing users to "sit in a proverbial director's chair and preside over 'auditions' at the same time as one can feel the process is 'less superficial' than other dating services."[56]

Tinder exemplifies the objectification of others by removing all social cues from the interaction between two users other than the single photograph, and reducing the time spent in deciding to the single moment in time required to swipe. Even the written profiles, which could provide more social cues than the simple image, take too long to read before deciding whether the user is interested in "dating" another user. All the social cues of

---

51. Turp, "Social Media, Interpersonal Relations and the Objective Attitude," 5.

52. Turp, "Social Media, Interpersonal Relations and the Objective Attitude," 6.

53. Turp, "Social Media, Interpersonal Relations and the Objective Attitude," 6.

54. David and Cambre, "Screened Intimacies," 2.

55. David and Cambre, "Screened Intimacies," 2.

56. David and Cambre, "Screened Intimacies," 6.

an initial romantic encounter are compressed into a single image judged in a fleeting moment in time.

Information Overload

When a speaker or entertainer stands in front of a large audience, the individual audience members "disappear into the crowd." There is not a person the speaker is performing for, but rather a mass of people, often barely visible through bright lights. Treating all the members of an audience as a single mass in this way is a form of flattening. In the case of the public speaker, the sheer amount of information pouring into their mind simply cannot be processed—there is no way to see the person in the fourteenth row falling asleep, the person in the twentieth row nodding vigorously, and the person in the front row who appears to want to ask a question, to process all of these things, and react appropriately to them.

Sherry Turkle argues that the sheer volume of information users encountered on Internet-based services poses this same sort of problem: "When you are besieged by thousands of emails, texts, and messages—more than you can respond to—demands become depersonalized."[57] According to *Lifewire*, 293 billion emails were sent (and received) each day in 2019; this is expected to grow to over 347 billion per day by 2023.[58] As the pace of communication increases, the amount of time any one person can spend on each individual communication decreases, exposing users to ever-smaller "slices" of other users in the system.

A person's inability to process the amount of information they receive each day creates a natural filter effect, reducing the information exchanged to the minimum necessary, and leaving out all the human components that expose a fully formed person. Turkle writes, "If when on the net, people feel just 'alive enough' to be 'maximizing machines' for emails and messages, they have been demoted."[59] It is easy to treat people in information-rich environments as simple information sources, rather than as individuals with dignity and value.

57. Turkle, *Alone Together*, 168.
58. Tschabitscher, "19 Fascinating Email Facts."
59. Turkle, *Alone Together*, 168.

Quantification

Critics of all stripes—film, art, television, theater, etc.—are widely known as a difficult audience. Anyone who must perform, whether by cooking a perfect meal or playing a piece to perfection—knows the sweaty palms, shortness of breath, and other physical symptoms brought on by the pressure of being measured by someone who knows how to measure, where to look to find mistakes, and has an audience who will listen to their critique.

On neurodigital systems, such as social media, everyone is a performer, and every post is a performance. The audience might not be well-educated, but it is vast, able to replay and examine the performance in minute detail forever, is able to alter the post to either mock or praise the performance, and every audience member may well have a large following to whom they can speak. Posting on neurodigital systems is like performing for an audience full of critics, all of whom have the tools to measure everything about your performance, an almost unlimited amount of time to consider the performance, and a vast audience to talk to.

The result of the measurement process might be a like, a comment, a reshare, or some other sign of approval. The result of the measurement process might be being cancelled, losing your job or your future, being called names, or many other things. Considering the kind of pressure the average teenager is under when performing in such a public space for such a critical audience, having so many tools and time at hand, is it any wonder heavy social media use can cause so many mental problems?

To make matters worse, these likes and reshares and comments are then quantified by counters and ratings in virtually every system built using neurodigital media. Quantification, according to Luis de Molina, is the assignment of value in the form of numeric value; for example, giving a loaf of bread a price is quantifying its value within a given context.[60] The process of quantification often leads to what Benjamin Grosser calls a culture focused on "the desire for more,"[61] creating an economy Devin Singh calls "endless growth."[62] Edd Noell says the quest for growth results in an "almost eschatological hope for a day in which the ongoing pursuit of material abundance is terminated and other significant, morally pure pursuits are enabled."[63]

60. Camacho and de Molina, "Treatise on Money," 185.
61. Grosser, "What Do Metrics Want?"
62. Singh, "Review," 59.
63. Noell, "An End to Scarcity?," 47.

According to Grosser, this quest for more transfers from the economic realm to the social realm through the quantification of approval and relationships. He writes,

> Quantification becomes the way we evaluate whether our desire for more is being fulfilled. If our numbers are rising, our desire is met; if not, it remains unmet. Personal worth becomes synonymous with quantity. Further, through strategies like the audit, the pursuit of capital establishes a desire to impress others through quantification that also plays into Bourdieu's ideas of symbolic and social capital. We want to "win" the confidence of our friends, to accumulate a capital of "social connections, honourability and respectability" that can be exchanged later within our social system. As we will see later on, when combined with metric visibility in Facebook, our desire for social capital leads us to internalize the need to excel in a quantified manner.[64]

The user's performance becomes driven by a desire to gain approval, turning other users into an abstract and flattened audience to be gained or a form of capital in a social network, rather than individuals having individual dignity to be respected and befriended.

Likes and reshares are not enough, however. Sharing applications, such as Uber, Lyft, and Airbnb also quantify relationships by allowing the direct rating of another user. What would once have been a taxi ride is turned into an (often) strained attempt for driver and passenger to treat one another as close friends—after all, your rating as a rider will determine if other drivers will pick you up, and your rating as a driver determines who else will hire you to drive.

Kari Paul describes staying in an Airbnb in Brazil, where her host "invited me to eat a traditional local snack of pão de queijo and have tea."[65] The problem was that the author was tired, and just wanted to go to bed—but felt like it was best to "play along." No point in being rated as a poor customer because you would not indulge in some local tradition. The result, however, is the quantification of every commercialized relationship—the user and the service provider both become performers on a larger stage, a stage which they cannot afford to be booed off.

The flattening, depersonalizing effect of all this performance and rating can hardly be overstated.

---

64. Grosser, "What Do Metrics Want?," 2.
65. Paul, "Rating Everything."

Flattening at Scale

Those involved in the field of information technology often say, "If the service is free, you are the product." Garson O'Toole traces this quote back through several variations to television advertising in the 1970s, but it is more valid of neurodigital media than it ever was of television. Zuboff describes Google's shift from using behavioral data for product improvement to using this data to satisfy the "burgeoning demand of advertisers eager to squeeze and scrape online behavior by any available means in the competition for market advantage."[66] The result, she says, is that "users were no longer ends in themselves but rather became the means to others' ends."[67] This is a form of flattening at scale.

One way in which operators flatten users is treating them as manipulable objects, clandestinely overriding their individual free will by nudging or manipulating users in directions perceived as economically or socially advantageous. However, this aspect of flattening will not be considered here because it was discussed in depth in chapter 2. The subsections below consider two other aspects of this flattening at scale: shaping information about users for economic gain and selling information about users for economic gain.

Shaping Information about Users for Economic Gain

Individual users are not the only ones on social media sites (and other systems based on neurodigital media) who tightly control what they post and what is posted about them—organizations, including the operators themselves, do the same thing.

Social media networks tightly control the images users see on their landing pages, advertisements, and even in user posts, to, for instance, make their services appear more attractive. According to Sam Biddle, Paulo Victor Ribeiro, and Tatiana Dias, TikTok moderators were instructed to de-emphasize images with "abnormal body shapes," "too many wrinkles," "eye disorders," and other low-quality traits so they would not be highlighted in users' feeds.[68] This is very similar to the techniques used by restaurants who place more attractive clients in window seats, so they will be seen by passersby, making the business appear more appealing to potential customers.[69]

66. Zuboff, *The Age of Surveillance Capitalism*, 88.
67. Zuboff, *The Age of Surveillance Capitalism*, 88.
68. Biddle, Ribeiro, and Dias, "Invisible Censorship."
69. Thompson, "What You Look Like."

Tinder's logic and interface perform the same filtering effects. Gaby David states, "Tinder's platform excludes users from freely defining how they will interact with others and domesticates them by shaping the social dynamics that in turn depend on the platform."[70] He argues this is accomplished by promoting some posts and not others, and by forcing users into a binary choice about other users, materially "influencing attitudes through speed and repetition."[71] Social media networks, then, flatten their users by reducing the presentation of the person to a small subset of the whole person.

Selling Information about Users for Economic Gain

According to Bruce Schneier, the initial noncommercial status of the Internet set an expectation that services offered through the Internet would be free. "Free," according to Schneier, is a "special price," warping "our normal sense of cost vs. benefit," and causing people to sell information about themselves for less than it is worth.[72] Once the initial expectation of "free" was set, commercial services found it impossible to charge anything for their content, driving them to conclude that "advertising was the only revenue model that made sense."[73] Targeted advertising is more effective than broad-based advertising, so targeted advertising quickly became the foundation for many of the commercial services carried through the Internet. The more targeted advertising is, the more effective it is; the ultimate goal is what Christopher Summers, Tobert Smith, and Rebecca Reczek call an "audience of one."[74]

To reach ever closer to this audience of one, service providers rely on nearly ubiquitous surveillance, described in more detail in the next chapter, to gather a massive amount of information. This information is then processed using digital technologies (as described in the chapter 1) to create what Zuboff calls behavioral surplus, which is then "fabricated into prediction products that anticipate what you will do now, soon, and later."[75] As with anything of value, Zuboff says a "marketplace for behavioral prediction" developed, which trades in what she calls "behavioral futures."[76]

70. David and Cambre, "Screened Intimacies," 9.
71. David and Cambre, "Screened Intimacies," 9.
72. Schneier, *Data and Goliath*, 49.
73. Schneier, *Data and Goliath*, 49.
74. Summers, Smith, and Reczek, "An Audience of One."
75. Zuboff, *The Age of Surveillance Capitalism*, 7.
76. Zuboff, *The Age of Surveillance Capitalism*, 7.

Jagdish Achara, Javier Parra-Arnau, and Claude Castelluccia describe the operation of the targeted advertising market:

> [A]dvertisers engage the services of an ad platform, which is responsible for displaying their ads on the publishers' sites. In this latter model, the ad-delivery process begins with publishers embedding in their sites a link to the ad platform they want to work with. The upshot is as follows: when a user retrieves one of those websites and loads it, their browser is immediately directed to all the embedded links. From these links, the ad platform may track this user's visit and display the ads of the advertisers it partners with. Within the third-party ad model, real-time bidding (RTB) is the dominant technology with 74% of programmatically purchased advertising.[77]

Many other services soon joined markets for targeted advertising for user data; Abbas Razaghpanah and colleagues identified 2,121 third-party tracking services operating in 2018.[78] The researchers say that "[d]ue to the opacity of the tracking ecosystem, it is difficult to uncover and track how organizations collect personal data from end-users, and how they store and share it with each other."[79]

The data analyzed and sold in this behavioral marketplace goes far beyond social media posts, personal networks, and searches. Ava Kofman notes Google's Sidewalk Labs proposed selling "cell phone location information 'in the aggregate' to help city planners"[80] in a smart city project. This information appears to be widely available even outside smart city projects; David Marcus notes cell phone location data was used to determine how many people obeyed the COVID-19 lockdown orders imposed by state and local governments in early 2020.[81] Kalev Leetaru describes pharmacies selling prescription information to data brokers, which is then combined through analysis with other data to drive targeted advertising of medical products.[82]

The result, according to a report by *MIT Technology Review*, is that "data is now a kind of capital, on par with financial and human capital in creating new digital products and services," eventuating in an economy

77. Achara, Parra-Arnau, and Castelluccia, "Fine-Grained Control."
78. Razaghpanah et al., "Apps, Trackers, Privacy, and Regulators," 1.
79. Razaghpanah et al., "Apps, Trackers, Privacy, and Regulators," 12.
80. Kofman, "Google's Sidewalk Labs."
81. Virus, "Despite Polling."
82. Leetaru, "How Data Brokers and Pharmacies Commercialize."

where "more important assets in the economy are composed of bits instead of atoms."[83] The report continues,

> To call data a kind of capital isn't metaphorical. It's literal. In economics, capital is a produced good, as opposed to a natural resource, that is necessary for the production of another good or service. Data capital is the recorded information necessary to produce a good or service. And it can have long term value just as physical assets, such as buildings and equipment, do.[84]

The status of data as capital can be shown through the many ways in which it is valued.

Will Rinehart states, "Most popular data valuations are accomplished through income derivations, often by simply dividing the total market capitalization or revenue of a firm by the total number of users."[85] This method leads to valuations such as $112 for the data of each user in the United States, or $260 per LinkedIn user in its purchase by Microsoft[86]—apparently low prices in terms of value for a person's total identity. Others argue these numbers are too high; Kugler says the value per user is often between "$15 and $40," and individual bits of information about a user being sold for as little as $0.00007.[87] Many of these valuations, however, miss the primary point—it is not the profile of the user or the individual data point that create value, but rather the way these things can be used to create and hold a relationship between the user and a brand (or idea), and the way they can be used to predict and modify future behavior.

As Zuboff says, "Surveillance capitalists discovered that the most predictive behavioral data come from intervening in the state of play in order to nudge, coax, tune, and herd behavior toward profitable outcomes."[88] She argues that the automation of the processes used to extract behavioral excess, through competitive pressure, eventually shifts the goals from "automat[ing] inflows about us; the goal is now to automate us."[89] The result is what Zuboff calls instrumentarianism, a condition where "power knows and shapes human behavior towards others' ends," working "its will through

83. MIT Technology Review Insights, "The Rise of Data Capital," 2.

84. MIT Technology Review Insights, "The Rise of Data Capital," 3.

85. Rinehart, "Hearing on Data Ownership," 4.

86. Rinehart, "Hearing on Data Ownership," 4.

87. Kugler, "The War Over the Value of Personal Data."

88. Zuboff, *The Age of Surveillance Capitalism*, 8.

89. Zuboff, *The Age of Surveillance Capitalism*, 8.

the automated medium of an increasingly ubiquitous computational architecture of 'smart' networked devices, things, and spaces."[90]

Conclusion

While previous chapters provided background, this is the first chapter to
really look at the impact of neurodigital media on individual users—this
chapter has probably been harder to digest than previous chapters. Understanding every nuance, however, is not as important as understanding the
overall picture: systems built using neurodigital media cause users to flatten
themselves and others. Flattening leads to objectification or treating other
persons—who should be treated as dignified ends—as mere means to an
end instead. To put this in plain terms: neurodigital media encourages users
to treat others as tools in a never-ending quest for a full ego. This ultimately
destroys the dignity of every user, because every user is both a performer
and a critic.

90. Zuboff, *The Age of Surveillance Capitalism*, 6.

7

Privacy

"IF YOU HAVE DONE nothing wrong, you have nothing to hide." Such is one of the most common sayings in our culture—and yet, one no one seems to really believe. Instead, this trite saying almost seems like a sigh of despair, interpreted as something like "we cannot do anything about massive surveillance, so we might as well find some way to rationalize it." Characterizing the scope of modern surveillance systems is difficult; there is no central authority, even in authoritarian regimes such as China, that keeps track of every surveillance program, how much information is collected, and how it is used.

Data breaches, however, can provide some sense of the scope of data being collected. For instance, in 2019 a researcher discovered an insecure database on the public Internet "filled with records tracking the location and personal information of 2.6 million people located in the Xinjiang Uyghur Autonomous Region. The records include individuals' national ID number, ethnicity, nationality, phone number, date of birth, home address, employer, and photos."[1] During one twenty-four-hour period, "6.7 million individual GPS coordinates were streamed to and collected by the database, linking individuals to various public camera streams and identification checkpoints associated with location tags such as 'hotel', 'mosque', and 'police station.'"[2]

In October of 2017, Yahoo exposed the personal information of 3 billion users. In 2018, Aadhaar, a biometric service provider, exposed personal information about 1.1 billion people, and Marriott exposed personal

1. O'Brien, "Massive Database Leak."
2. O'Brien, "Massive Database Leak."

114

information about 500 million users. In 2019, First American Financial exposed personal information about 885 million users; an email verification service exposed personal information about 763 million users; Facebook exposed personal information about 540 million users; and Yahoo exposed personal information about 500 million users.[3]

Smaller breaches happen almost daily but are not reported simply because they are so small. A few thousand records here or there are almost of no consequence—except to the people whose identities are stolen, whose credit cards are used for fraudulent purposes, and whose lives are ruined. Larger breaches are reported in the technical press, and sometimes picked up by more widely read news outlets, but the numbers are so large—and the details so scanty—it is almost impossible to process what happened.

These kinds of databases exist in every nation. Peter Waldman, Lizette Chapman, and Jordan Robertson published an article in 2018 about a company called Palantir, which began as a private surveillance company just after the terrorist attacks on September 11, 2011, in the United States.[4] According to Waldman and coauthors, Palantir quickly became the choice of large financial organizations to provide internal employee security, "vacuum[ing] up emails and browser histories, GPS locations from company-issued smartphones, printer and download activity, and transcripts of digitally recorded phone conversations."[5] Palantir's clients have since grown to include government agencies, such as the Los Angeles Police Department.[6]

Why is this important? While some scholars argue privacy simply is not important or is a recent cultural innovation,[7] the argument that privacy is important to human dignity is strong, particularly from within a Christian view of the person.

Where does all this information come from? First, users of the almost pervasive neurodigital media systems simply give it to the operators in trade for a bit of convenience or as part of a performance. Users simply do not understand the amount of information they give away. Second, information comes through pervasive surveillance—to levels unknown in history.

3. Tunggal, "The 36 Biggest Data Breaches."
4. Waldman, Chapman, and Robertson, "Palantir Knows Everything About You."
5. Waldman, Chapman, and Robertson, "Palantir Knows Everything About You."
6. Waldman, Chapman, and Robertson, "Palantir Knows Everything About You."
7. See for instance Gotlieb, "Privacy."

The Importance of Privacy

This wholesale and pervasive collection of information about individuals, and the further intrusion into individuals' lives created by the inferences drawn from this information, leave the individual user with little or no privacy from the operator's gaze. The importance of privacy to human dignity, however, can hardly be overstated.

In Genesis 9:22, "Ham, the father of Canaan, saw the nakedness of his father and told his two brothers outside." According to Nahum Sarna, the exposure of Noah in this way "is associated with shame and with loss of human dignity, as Genesis 3:7, 21 make clear," particularly as this was an invasion of Noah's privacy.[8] The curse Noah places on Ham and his descendants in Genesis 3:25–27 testifies to the seriousness of the loss of privacy and its connection to dignity in the scriptural record. Ham's invasion of Noah's privacy also impacts the autonomy of Noah and Ham. Noah's autonomy is reduced because Ham now "knows" more about Noah than he would otherwise choose to reveal. Ham's autonomy, in return, is reduced because of Noah's curse.

Sarna, in his commentary on Genesis 9:22, refers to Genesis 3:7, which describes the fall of humanity. On disobeying God's commandment not to eat the fruit of the tree of knowledge, Adam and Eve discover they are naked, and try to cover themselves with leaves. This is not, however, successful. God, at this point, could have told the couple they must live in the shame of exposure—but he does not. Instead, God makes them effective clothes, ones that will truly cover their bodies, covering their shame. Clothing, from this point forward, is regularly tied to the covering of shame. The garments the priests wore in the temple were for their honor (Exodus 28:1), and the steps of the temple were to not be too steep so the nakedness of those ascending them would not be uncovered.

There are many other instances of privacy in the Scriptures. Gossip, or talking about the private affairs of others, is often condemned. Proverbs 11:13 condemns those who "reveal secrets," and praises those who "keep a thing covered." Jesus often went off alone to pray (Luke 6:12) and instructed his followers to pray in private as well (Matthew 6:5). Privacy is clearly not a foreign concept in the Scriptures.

Nor is privacy treated lightly among secular thinkers. Privacy, according to Alan Westin and Daniel Solove, is "the voluntary and temporary withdrawal of a person from the general society through physical or psychological means, either in a state of solitude or small-group intimacy or,

8. Sarna, *Genesis*, 65.

when among larger groups, in a condition of anonymity or reserve."[9] Daniel Solove says there is a "worldwide" consensus that privacy is a "fundamental human right," based on its inclusion in the United Nations Universal Declaration of Human Rights of 1948 and the European Convention of Human Rights of 1950.[10] Many, however, seem to believe individual privacy is not something to be sought—often expressed in a saying such as: "if you have nothing to hide, you have nothing to fear." Richard Epstein says, "The plea for privacy is often a plea for the right to misrepresent one's self to the rest of the world."[11] This sentiment, however, illustrates a crucial mistake in understanding the human need for privacy.

Westin and Solove argue that privacy is linked to autonomy because the "deliberate penetration of the individual's protective shell, his psychological armor, would leave him naked to ridicule and shame and would put him under the control of those who knew his secrets,"[12] even if that penetration can be justified. Westin and Solove consider the autonomy gained through the privacy of hiding information, which may be used to manipulate an individual in terms of his "basic dignity and worth as a creature of God and a human being, and in the need to maintain social processes that safeguard his sacred individuality."[13] They note that all individuals need some time to "lay aside their masks," and "to always be 'on' would destroy the human organism."[14]

The autonomy privacy affords, according to Westin and Solove, is critical to the development of individuality, because privacy gives each person the time and space to "integrate his experiences into a meaningful pattern and to exert his individuality on events."[15] Elizabeth Corey agrees with this assessment, saying,

> Radical self-exposure leaves no place for the privacy necessary to form an authentic identity, as opposed to a performative one. Human beings need a realm in which we are free to act without anyone watching, without wondering what our recollection of the moment will look like on Facebook, and without having to produce some witty remark that will show how worldly-wise we are. Privacy allows us to engage in activities for their own sake.

9. Westin and Solove, *Privacy and Freedom*, loc. 294.

10. Solove, *Understanding Privacy*, loc. 64.

11. Epstein, "The Legal Regulation of Genetic Discrimination," 12.

12. Westin and Solove, *Privacy and Freedom*, loc. 855.

13. Westin and Solove, *Privacy and Freedom*, loc. 855.

14. Westin and Solove, *Privacy and Freedom*, loc. 919.

15. Westin and Solove, *Privacy and Freedom*, loc. 948.

It also allows us to be sincere without embarrassment and to act without wondering how others may evaluate us.[16]

Corey likens this need for privacy to the confessional seal of the Roman Catholic tradition, an "impenetrable zone of privacy" within which each person feels "free to expose the secrets of our souls."[17] Westin and Solove call this the "phenomenon of the stranger," a situation where pieces of information about oneself that would not be shared with even the closest friend or confidant, will be shared with a stranger because the stranger is "able to exert no authority or restraint over the individual."[18] In the case of neurodigital media, there is no space for the "phenomenon of the stranger" to take place because the system itself records every act of communication for later retrieval and analysis.

Privacy is directly tied to identity construction—as Adam Moore says, it "is a necessary condition for human well-being or flourishing and not a mere interest."[19] In overriding the development of the autonomous, morally free, individual, pervasive surveillance undermines the dignity of individual humans.

The Extraction Imperative, Data Gravity, and Curiosity

The veil of privacy so needed to support identity formation and the continued operation of the social order, however, is regularly pierced by the pervasive surveillance needed to create the behavioral surplus that makes neurodigital media possible. Operators justify these pervasive surveillance operations primarily by citing the economic and social goods produced by risk-making.

While risk is most often associated with something that may happen in the future, risk-making is a process of predicting future events within a range of probability. Hence, risk itself, according to Dan Bouk, becomes a kind of commodity.[20] Commodifying the future by exposing the probability of any particular future taking place allows planning, which allows manufacturers to decide which product to make, how much of each product to make, and when it will be desired.

16. Corey, "Our Need for Privacy," 48.

17. Corey, "Our Need for Privacy," 48.

18. Westin and Solove, *Privacy and Freedom*, loc. 836.

19. Moore, "Privacy, Speech, and Values," 41.

20. Bouk, *How Our Days Became Numbered*, loc. 196.

These predictions are at the core of just-in-time delivery, described by Caroline Banton as a system where the materials and products needed to produce and sell a product are delivered just when they are needed, rather than being delivered early and stored for indeterminate amounts of time before being used.[21] Eric Stoddart writes,

> Surveillance is built upon both the construction of risk and its statistical analysis; although the boundaries become blurred as we shall see. Risk of robbery or of insurance fraud motivates commercial interests to, understandably, limit their losses. Medical surveillance across, as well as within, national borders relies on mathematical modelling of the spread, intensity and outcomes of disease whether that be of HIV in parts of Africa, H_1N_1 avian flu virus across the world or more localised occurrences of health problems. States such as China who pursue strategies to keep its citizens' Internet use under surveillance are also concerned with risk—not the probabilistic, actuarial kind, but their perception and evaluation of the danger posed by dissidents when information control is loosened in an authoritarian regime. The overhead cameras that monitor traffic flows on most highways and at strategic junctions in our major cities are there to speed movement and quickly identify accidents, but with the spectre of economic and environmental costs within a capitalist system that knows about the ever-present risks to investment and profit.[22]

Risk-making of this kind has quickly moved beyond the production and movement of physical goods and limiting losses. For instance, the Pretrial Justice Institute outlines how such risk-making is used in setting bail for criminal defendants,[23] and Hari Nair describes how these same risk-making tools can be used to determine when an employee might leave their position.[24]

Once using the risk-making capabilities of surveillance, analytics, and control—what Zuboff calls a "unique logic of accumulation"—has been justified, it is then solidified into a process that can "transform investment into profit."[25] The market for profitable, actionable information drives the accumulation of ever more significant amounts of data by surveilling

---

21. Banton, "Understanding Just-in-Time (JIT) Inventory Systems."
22. Stoddart, *Theological Perspectives on a Surveillance Society*, 4.
23. Pretrial Justice Institute, "Risk Assessment."
24. Nair, "A Predictive Analytics Model."
25. Zuboff, *The Age of Surveillance Capitalism*, 52.

ever greater parts of life. Zuboff calls this the extraction imperative, saying, "Industrial capitalism had demanded economies of scale in production in order to achieve high throughput combined with low unit cost. In contrast, surveillance capitalism demands economies of scale in the extraction of behavioral surplus."[26]

Within the information technology field, increasing the scope and intrusiveness of surveillance is driven by data gravity, one aspect of which is best described by Moshe Vardi as Kai-Fu Lee's virtuous cycle: "More data begets more users and profit, which begets more usage and data."[27] Arguments such as those made by Marijn Sax related to the ethics of dividing users from their data and the historical conception of justice about the extraction and use of data through surveillance to create behavioral excess are generally ignored in information technology circles.[28]

Two final reasons—often overlooked—for the constant expansion of surveillance are curiosity and a desire to enforce social norms. Curiosity can drive users to constant social comparison, as described earlier in this book. Curiosity, according to Emily Thomas, can easily turn into cyberstalking, an unhealthy obsession with another person pursued through digital media.[29]

Curiosity can also turn into a desire to enforce social norms; all societies have taboos that every person within the culture must strive to maintain. These norms are upheld through a combination of "socially approved machinery" by authorities, as noted by Westin and Solove,[30] and through the actions of individuals who invade the privacy of others to guard against antisocial conduct (such as the so-called Karens who invaded the privacy of others to report their breaching lockdown protocols during the COVID-19 pandemic).[31]

Pervasive Surveillance Collection Practices

Massive databases of information must come from somewhere. Who collects this information, and how is it collected? The scale and scope of data collection is difficult to fathom in its entirety; this section will give some examples of the many ways in which data is collected and used.

---

26. Zuboff, *The Age of Surveillance Capitalism*, 87.
27. Vardi, "The Winner-Takes-All Tech Corporation," 7.
28. Sax, "Big Data."
29. Thomas, "Cyber-Stalking."
30. Westin and Solove, *Privacy and Freedom*, loc. 582.
31. Borysenko, "The Ultimate Crowdsourced Definition of a Karen."

Personal electronic devices, such as cellular telephones, are the most common way data is collected on individuals. The easiest way to track the location of a phone is through its interaction with the Global Positioning System (GPS). Mahesh Balakrishnan, Iqbal Mohomed, and Venugopalan Ramasubramanian also describe a system of determining the (somewhat imprecise) location of a phone using its network address;[32] Kateřina Dufková and company describe a more accurate method using the phone's proximity to radio transmitters used to provide cellular telephone service.[33] More recently, Colleen Josephson and Yan Shvartzshnaider describe a new electronic chip added to Apple's iPhones that allow accurate location tracking relative to other Apple devices.[34]

The apps loaded on these phones are also a rich source of information. Exposing the network traffic created by apps loaded onto smartphones reveals that over 4,000 of these apps send data to at least one Advertising and Tracking System (ATS). Alphabet, Google's parent company, provides tracking services to close to 100 percent of the apps available in the Android ecosystem.[35] The Norwegian Consumer Council commissioned a study to determine which apps shared information with advertising services, as well. The researchers found virtually every app shares information with Google and Facebook, and most apps share information with three or more tracking services.[36]

Stephen Wicker and Dipayan Ghosh have also identified surveillance in electronic reading applications and devices, such as Amazon's Kindle.[37] Clifford Lynch documents a broader collection of reading through all forms of electronic reading.[38] According to Rachel Metz, Amazon's Alexa, an artificially intelligent voice-driven assistant that runs on physical items as diverse as clocks and cellular telephones, records every conversation within the range of its microphone.[39] One study researched the data transmitted by smart televisions and "set-top boxes" that connect televisions to the

---

32. Balakrishnan, Mohomed, and Ramasubramanian, "Where's That Phone?"

33. Dufková et al., "Active GSM Cell-ID Tracking."

34. Josephonson and Shvartzshnaider, "Every Move You Make."

35. Razaghpanah et al., "Apps, Trackers, Privacy, and Regulators," 7.

36. Kaldestad, "Out of Control: A Review." The parallel report Kaldestad, "Out of Control: How Consumers," provides a good overview of the advertising industry and how it uses private information for financial gain.

37. Wicker and Ghosh, "Reading in the Panopticon."

38. Lynch, "The Rise of Reading Analytics."

39. Metz, "Yes, Alexa Is Recording Mundane Details of Your Life."

provider's network, and discovered these devices also gather and record information about their users and their immediate environment.[40]

Specialized websites are another rich source of information on individual users. Timothy Libert, in a special review article, shows that "over 90% of the 80,142 health-related web pages" send information to other services, presumably for information tracking. The information these sites is sending goes far beyond just who the user is and where they are located; according to Libert, 70 percent of these health sites include information about specific symptoms, diseases, and treatments the user has examined.[41] The health data leaked to these sites when a user searches for information on a particular condition can be used by insurance companies, employers, government agencies, and pharmaceutical companies to determine coverage, make hiring (or firing) decisions, and drive drug marketing efforts at an individual level.

Brandon Murrill and Edward Liu expose information collection through electric meters, saying, "Detailed electricity usage data offers a window into the lives of people inside of a home by revealing what individual appliances they are using, and the transmission of the data potentially subjects this information to interception or theft by unauthorized third parties or hackers."[42] Many of the devices in homes are part of what is called the Internet of Things; Jingjing Ren and coresearchers have documented the number of information devices with embedded network connections report to centralized databases.[43]

Tracking is becoming common even in outdoor environments, where individuals might have, in the past, expected to have some amount of privacy. Adam Schwartz notes the City of Chicago "probably has access to somewhere between 10,000 and 20,000 publicly and privately owned surveillance cameras," which can record and observe the entire core of the city.[44] The city, according to Schwartz, does not release any information about the effectiveness of these cameras in preventing or solving crimes.[45] Most cities have similar systems in place, including cameras positioned to capture license plates for toll collection and traffic signal enforcement. Cameras are not the

---

40. Mohajeri Moghaddam et al., "Watching You Watch."

41. Henry, "The Jurisprudence of Dignity," 68.

42. Murrill and Liu, "Smart Meter Data," summary.

43. Ren et al., "Information Exposure from Consumer IOT Devices."

44. Schwartz, "Chicago's Video Surveillance Cameras," 47.

45. Schwartz, "Chicago's Video Surveillance Cameras," 50.

only way cities can surveil individuals; however, Jesse Marx describes the increasing surveillance capabilities of the smart streetlights in San Diego.[46]

In effect, there are now few places an individual can go or things an individual can do that will not produce data that is recorded, analyzed, and used in some system based on neurodigital media.

Combining Data Across Systems to Invade Privacy

Almost limitless inferences about individuals can be made from this veritable sea of data. The first and most straightforward of these is the location and identity of any individual. Beyond the presence of a personal device, one of the most common ways to identify a person is through facial recognition. In a 2016 report published by *Georgetown Law*, Clare Garvie and coauthors call the widespread use of facial recognition by law enforcement in the United States "the perpetual lineup."[47]

According to them, "law enforcement face recognition networks include over 117 million American adults,"[48] at the time of their report. One of the dangers they report with these uses of facial recognition is that "most law enforcement agencies do little to ensure that their systems are accurate."[49] Logan Kugler notes that in more modern systems, such as those used to screen passengers at airports, facial recognition is combined with voice recognition to produce a more accurate match.[50] These newer systems combine data from the images posted to social media services, such as Facebook, with the voice information gathered from personal voice assistants, such as Amazon's Alexa, to create a rich database of recognition data.[51]

There are many other types of location tracking and identification available to the designer of systems based on neurodigital media. Jingyu Hua, Zhenyu Shen, and Sheng Zhong describe a system that uses the accelerometer data available from a cellular telephone to find out where the person carrying the phone has traveled through a subway system even though cellular service and GPS location data is not available.[52] Their system uses the speed at which the phone is moving, combined with the known time between each stop, to infer when a user gets on a particular train, when

46. Marx, "The Mission Creep of Smart Streetlights."
47. Garvie et al., "Unregulated Police Face Recognition in America."
48. Garvie et al., "Unregulated Police Face Recognition in America," 2.
49. Garvie et al., "Unregulated Police Face Recognition in America," 3.
50. Kugler, "Being Recognized Everywhere."
51. Kugler, "Being Recognized Everywhere."
52. Hua, Shen, and Zhong, "We Can Track You If You Take the Metro."

they get off, where they change trains, and other information about their travels.[53]

Keith Kirkpatrick describes technologies used to track customers through a retail environment, such as a Bluetooth beacon that connects to each individual's cellular telephone, identifying each person, and tracking their progress through space.[54] The information gathered from this kind of tracking can be combined with the individual's browsing and purchasing history even when they do not have the store's application loaded on their phone using a side-channel identification, such as the one described by Daniel Arp and coauthors.[55] To create such a system, some unique identifier from an individual's phone would be learned and stored while the person is in the physical location shopping. Once the individual returns home, they would set their phone next to their computer. When the individual opens the retailer's website, the computer speaker can be used to send data encoded as a very high-frequency sound to an application on the phone, which can associate the phone, computer, and individual user. This association allows the information stored in the database about the movement of the phone to be related to the individual browsing the retailer's website, connecting the physical shopping experience to the virtual one.

Autonomy, Privacy, and Permission

A common counter to the problem of pervasive surveillance is that it only takes place with the permission of individual users. In this view, individual humans are exercising their autonomy by making a rational trade-off between privacy and convenience—or, in other terms, users of these services are allowing themselves to be flattened and objectified through autonomously exercised choices.

As Solove says, "Although polls indicate that people care deeply about privacy, people routinely give out their personal information and willingly reveal intimate details about their lives on the Internet."[56] Or, as Calvin Gotlieb declares, "Most people, when other interests are at stake, do not care enough about privacy to value it."[57] Three responses to this argument are considered in the sections below.

53. Hua, Shen, and Zhong, "We Can Track You If You Take the Metro."
54. Kirkpatrick, "Tracking Shoppers," 19.
55. Arp et al., "Privacy Threats."
56. Solove, *Understanding Privacy*, loc. 79.
57. Gotlieb, "Privacy," 156.

The Effectiveness of Privacy Controls

Some argue privacy should not be a concern when using large-scale systems built using neurodigital media because operators employ effective privacy controls. For instance, according to Facebook's privacy policy,

> We provide aggregated statistics and insights that help people and businesses understand how people are engaging with their posts, listings, Pages, videos and other content on and off the Facebook Products. For example, Page admins and Instagram business profiles receive information about the number of people or accounts who viewed, reacted to, or commented on their posts, as well as aggregate demographic and other information that helps them understand interactions with their Page or account.[58]

Aggregation is a common technique used by operators to preserve the privacy of their users. For instance, Facebook may reveal how many people have clicked on or viewed an update or an advertisement in a given time, rather than providing a list of individuals who have done so. Orange and some mobile telephone service providers in China release the number of users within a geographic region,[59] and Twitter releases information about the number of tweets regarding a topic along with some metadata (information about the tweets) for researchers and corporate partners to use.[60]

However, data aggregation is not effective at protecting user privacy. Simon Garfinkel, John Abowd, and Christian Martindale describe the reconstruction of private information through database reconstruction attacks.[61] These attacks use Boolean satisfiability problem (SAT) solvers. According to Martin Hořeňovský, these solvers find a set of data that will produce the same aggregated output as a given data set.[62] The computational complexity of these solvers has restricted their use in the past, but increasing computing power and optimized heuristics have made them tractable. Garfinkel and coauthors provide an example of using a SAT solver to recover personal information from an example set of aggregated census data.

Beatrice Perez, Mirco Musolesi, and Gianluca Stringhini show the metadata, or data about the data, released with aggregated information

---

58. Facebook, "Data Policy."

59. Xu et al., "Trajectory Recovery from Ash," 1241.

60. Perez, Musolesi, and Stringhini, "You Are Your Metadata," 1.

61. Garfinkel, Abowd, and Martindale, "Understanding Database Reconstruction Attacks on Public Data."

62. Hořeňovský, "Modern SAT Solvers."

by Twitter, provides enough information to correctly identify individual users, matching them with their portion of the aggregated information.[63] Perez and coauthors show how these tactics rely on seemingly innocuous information, such as the date and time the user's account was created, the number of tweets "favorited," the account's followers, the account's friends, and the number of public lists that include this account contain enough information to identify an individual user.[64] These methods use aggregated information publicly released by cellular telephone service providers to identify individual users.[65] Researchers determine that the only question they need to answer is, "are two masked mobility records created by a single user or different users?" They use three key facts to answer this question: users tend to have low mobility at night (when most users are sleeping), user's mobility tends to be continuous when they are moving (during the day), and each user's mobility pattern is self-consistent and generally unique from that of other users.[66] Once they have identified individual users through these three measures, they have found they can quickly identify a single user and trace their location across time.[67]

Database reconstruction techniques are complex; it is much simpler for application developers to take the information they want directly. The Android operating system used on many personal electronic devices, such as cellular telephones and televisions, regulates access to sensitive data through a privacy subsystem. Developers can access information such as the location of the device, the state of network connections (such as WiFi), the ability to send and receive messages, access to the public Internet, and many others.[68] Developers are supposed to ask permission from users before accessing any of this information, and they are supposed to abide by the choices users make about what information an app can access. Joel Reardon, Álvaro Feal, and Primal Wijesekera tested "hundreds of thousands" of apps in an instrumented environment to determine if these permissions were being followed or circumvented.[69] The researchers found "a number of side and covert channels that are being used to circumvent the Android permissions system," potentially impacting hundreds of millions of users.[70]

63. Perez, Musolesi, and Stringhini, "You Are Your Metadata," 1.

64. Perez, Musolesi, and Stringhini, "You Are Your Metadata," 3.

65. Xu et al., "Trajectory Recovery from Ash."

66. Xu et al., "Trajectory Recovery from Ash," 1–2.

67. Xu et al., "Trajectory Recovery from Ash," 6.

68. Android Developers, "Manifest.Permission."

69. Reardon, Feal, and Wijesekera, "50 Ways to Leak Your Data."

70. Reardon, Feal, and Wijesekera, "50 Ways to Leak Your Data."

The Norwegian Consumer Council funded a study using similar techniques to discover the amount of information transmitted by apps running on Android devices—whether these apps were disregarding user permissions was not considered. This study found that "ten apps were observed transmitting user data to at least 135 different third parties involved in advertising and/or behavioural profiling," and "[t]he Android Advertising ID, which allows companies to track consumers across different services, was transferred to at least 70 different third parties involved in advertising and/or profiling."[71] The average user probably does not understand the sheer volume of information shared by these devices and applications. Privacy controls, such as aggregation and user privacy controls, are not generally effective at controlling the flow of personal information from individual users to operators of systems built on neurodigital media.

The Trade-Off is Not Well-Informed

The first response is that the decision by users to give up privacy in neurodigital media is not fully informed. First, as Bruce Schneier points out, "This tendency to undervalue privacy is exacerbated by companies deliberately making sure that privacy is not salient to users."[72] An individual using social media networks, digital voice assistants (such as Amazon's Alexa), smart televisions, and other sensors in the pervasive surveillance of neurodigital media will find all the default permissions set to "share as much information as possible"—what Mark Zuckerberg calls "frictionless sharing."[73] If a user happens to search for intentionally or accidentally stumble across a services privacy statement, Mika Klaus says they will find it too "complex, too long, and consumers often do not have knowledge about the company's information practices after reading the long notices."[74] Privacy becomes incidental, rather than central, to using the service.

Second, as Zuboff notes, the language used to describe what these operators are collecting is often deceptive:

> Two popular terms—"digital exhaust" and "digital breadcrumbs"—connote worthless waste: leftovers lying around for the taking. Why allow exhaust to drift in the atmosphere when it can be recycled into useful data? Who would think to call such

71. Kaldestad, "Out of Control: A Review"; and Kaldestad, "Out of Control: How Consumers."

72. Schneier, *Data and Goliath*, 49.

73. As quoted by Manjoo, "Facebook's Terrible Plan."

74. Klaus, "Privacy Statements."

recycling an act of exploitation, expropriation, or plunder? Who would dare to redefine "digital exhaust" as booty or contraband, or imagine that Google had learned how to purposefully construct that so-called "exhaust" with its methods, apparatus, and data structures? The word "targeted" is another euphemism. It evokes notions of precision, efficiency, and competence. Who would guess that targeting conceals a new political equation in which Google's concentrations of computational power brush aside users' decision rights as easily as King Kong might shoo away an ant, all accomplished offstage where no one can see?[75]

Julie Cohen says the average user believes operators are collecting what "appears to be the ultimate disembodied good, yielding itself seamlessly to abstract, rational analysis."[76] Because information is a disembodied good, having little to do with the person or the person's physical being, it is hard to raise serious objections to its collection and use.

Finally, Westin and Solove note, "Man still relies heavily on his 'animal' or physical senses—touch, taste, smell, sight, and hearing—to define his daily boundaries of privacy. What is considered 'too close' a contact and therefore an 'invasion of privacy' in human society will often be an odor, a noise, a visual intrusion, or a touch; the mechanism for defining privacy in these situations is sensory."[77] It is difficult to see how individuals are making well-informed decisions about the collection and use of their data—the pervasive invasion of their privacy—when they do not understand the nature of what they are trading and are not mentally equipped to understand where the limits to this invasion should be placed.

Hiding the full consequences of trading privacy for convenience or some other good is often a form of dark pattern. Operators obfuscate the choice between privacy and some other good by placing the decision within a choice architecture favoring the operator's best interests, rather than the individual user—nudging the user toward making a decision that is not in their best interests. Beyond the direct harm to dignity through the loss of privacy, creating a choice architecture that nudges users towards making decisions that are not in their best interests harms the autonomy of individuals.

---

75. Zuboff, *The Age of Surveillance Capitalism*, 90.
76. Cohen, *Configuring the Networked Self*, 20.
77. Westin and Solove, *Privacy and Freedom*, loc. 338.

The Powerlessness of Users

A counter to the argument that individuals are making a rational, autonomous decision between privacy and convenience is that most users feel powerless to preserve their privacy. Joseph Turow, Michael Hennessy, and Nora Draper undertook a large-scale project to understand why American consumers claim to care about their privacy, and yet seem to be ready to give away information about themselves to gain some convenience. They conclude that "a majority of Americans are resigned to giving up their data—and that is why many appear to be engaging in tradeoffs."[78] They continue, "By misrepresenting the American people and championing the tradeoff argument, marketers give policymakers false justifications for allowing the collection and use of all kinds of consumer data often in ways that the public find objectionable."[79]

Alessandro Acquisti, Curtis Taylor, and Liad Wagman argue many individuals have simply given up on protecting information about themselves because even if they do not disclose anything, others will.

> [E]ven an individual's costs (and ability) to protect her information may be a function of the disclosure choices made by others. That "anonymity loves crowds" is a common refrain in the literature on privacy-enhancing technologies, reflecting the observation that, online as offline, it is easier to hide as one among many who look alike. Conversely, protecting one's data becomes increasingly costly the more others reveal about themselves (for instance, the success of online social networks has encouraged other entities, such as online news sites, to require social media user IDs in order to enjoy some of their services, thus curtailing users who do not want to create social media accounts), or altogether infeasible.[80]

This resignation is understandable once the partnerships between government entities and large neurodigital media platform providers are considered, such as Google's Sidewalk Labs project, which plays a role as an "Innovation and Funding Partner, including a role as lead developer of real estate," and provides technology to help a city with managing energy use and people.[81]

78. Turow, Hennessy, and Draper, "The Tradeoff Fallacy," 3.
79. Turow, Hennessy, and Draper, "The Tradeoff Fallacy," 3.
80. Acquisti, Taylor, and Wagman, "The Economics of Privacy," 446.
81. Jaffe, "Toronto Tomorrow," 39.

Conclusion

Privacy is necessary for human flourishing and demanded by human dignity—both from a Christian and secular standpoint. Yet surveillance is so common it is almost impossible for the average person to do anything without being tracked. As contact tracing is deployed to counter pandemics, more commerce moves online, higher percentages of people work from home, cars and other "things" in our lives become more connected, smart electric meters and security cameras become more common, and other forms of surveillance proliferate, the problem of surveillance is going to get worse rather than better.

Perhaps one of the most frustrating problems with surveillance is how little the average user believes they can do about it. Surveillance by private companies is not well-controlled by governments—at least in part because governments want the information generated—so individual users cannot seem to rely on governments to solve this problem. In the case of government-led programs, private companies are often used as "shields" for any violation of the relevant laws through a public-private partnership.

There are things individuals can do to reduce the impact of surveillance in their lives, such as simply turning their smartphones off in certain situations, but a full discussion of these practices is outside the scope of this book. They will be covered in more detail in a future book, however.

8

Freedom

DIGITAL SYSTEMS, FROM SOCIAL media to search engines, are supposed to free their users from the strain of everyday reality. Larry Page, cofounder of Google, says, "Rapid improvements in artificial intelligence, for instance, will make computers and robots adept at most jobs. Given the chance to give up work, nine out of 10 people 'wouldn't want to be doing what they're doing today.'"[1] One particular matter of concern for those promoting digital technology is the amount of time and effort required for humans to make decisions.

Much of the work that goes into replacing human decision-making is predicated on the belief, as Robert Cringley says, that "we have ancient brains that generally don't do the math because math has only been around for 5,000 years or so while we as a species have been walking the Earth for 2+ million years. Our decision-making processes, such as they are, are based on maximizing survival not success."[2] So the kind of decision-making humans excel at do not, however, match the real world any longer. Instead, now that we have vast amounts of information available about everything from apples to people, what is needed is the ability to crunch numbers, look for patterns, and make decisions based on those patterns.

Cringley argues one place where using machines to make decisions is taken seriously is Google:

> At Google they do what the algorithm tells them to do. So the algorithm is, itself, in charge until enough time passes that a

1. Waters, "FT Interview."
2. Cringley, "Welcome to the Post-Decision Age."

preponderance of data makes it clear the algorithm has failed. But even then they don't reject the algorithmic approach, they just revise the algorithm. This, I believe, is the trend. As humans we're pretty much all instinct shooters but the optimization of complex systems requires sight shooters. If we can't become those we use machines to do the work. And if the machines fail we don't reject them, we improve them.[3]

Google takes the blending of human and computer decision-making so seriously it extends the discipline of decision science to include engineering and interaction with deep learning networks.[4]

The human cost of this revolution in decision-making is much more than simply being managed by a computer—handing decisions to computers ultimately harms the moral freedom of users. This chapter opens by explaining how we came to accept machines making and influencing decisions. With this background in place, the nudge is described—the least intrusive (and yet, in many ways, the most effective) way to model and modify human behavior. This chapter then moves to habits, and then addiction. Finally, the moral dimensions of computers making decisions for people will be considered.

Accepting the Machine as Decision Maker

Trust in digital systems has reached heights previous generations could not have imagined. Computers have replaced human managers in many places,[5] determining how long employees will work and what they will do. People will even follow robots in emergency situations even when the robot is obviously not leading them to safety or is doing something that will endanger their lives.[6] How did we come to the point of placing so much trust in machines to make our decisions?

Seeking Statistical Certainty: Life Insurance

Beginning in the 1830s and 1840s, companies sold life insurance policies that would replace the income of the household head (normally the father) with a cash payout in the case of death. According to Dan Bouk,

3. Cringley, "Welcome to the Post-Decision Age."
4. Byrne, "Why Google Defined a New Discipline."
5. Dzieza, "Robots Aren't Taking Our Jobs—They're Becoming Our Bosses."
6. Wagner, Borenstein, and Howard, "Overtrust in the Robotic Age."

life insurance companies could sell such insurance by "turn[ing] a life into money."[7] To be profitable, companies could not pay out more in claims than they received in insurance payments, so they sought ways to determine if a person was a good risk—were they financially stable (able to pay the premiums indicated), physically healthy, and not likely to commit suicide (they would pay premiums long enough to cover any eventual payout on the part of the insurance company)? Bouk says that "in the bargain for economic security, insurance applicants allowed themselves to be made into commodified risks,"[8] a process that paralleled the commoditization of many other natural resources during the same time period.[9] This risk-making, however, soon turned to helping people live better lives.

According to Bouk, the leaders of the most influential life insurance companies in the 1900s started to ask why they should limit themselves to forecasting the future "when risk-makers' methods and tools—built for statistical prediction—could also change the future."[10] If the use of statistics could determine that a person's life would probably be shortened for some reason, it should be possible to extend their life by eliminating or controlling whatever was shortening their lives. For instance, doctors could encourage people to maintain their weight within a range shown by statistical analysis to be correlated with longevity and quality of life. Bouk argues that insurance executives saw that helping customers increase the quality and length of their lives would increase profits, would show insurance companies cared about those with whom they were doing business, and would improve society.[11] Predicting correlations between health factors and results required insurance companies to gather ever increasing amounts of personal data about their customers.

Encouraging (or even requiring) yearly doctor's visits not only allowed insurance companies to gather the information they needed, it also put individual consumers in regular contact with doctors who could shape behavior towards more optimal outcomes. Insurance companies funded the creation of nonprofit organizations that would encourage yearly doctor's visits and the gathering of health data they needed, such as the Life Extension Institute.[12] By gathering personal information in order to better predict risk, life insurance companies normalized trading personal information for

7. Bouk, *How Our Days Became Numbered*, preface.
8. Bouk, *How Our Days Became Numbered*, preface.
9. Bouk, *How Our Days Became Numbered*, preface.
10. Bouk, *How Our Days Became Numbered*, loc. 115.
11. Bouk, *How Our Days Became Numbered*, loc. 121.
12. Bouk, *How Our Days Became Numbered*, loc. 131–32.

financial and personal benefit—that is, those who gave their information to life insurance producers could obtain lower rates and gain insight on how to live a longer, healthier life. By requiring their customers to see a doctor yearly and accept advice on how to be healthy, life insurance companies normalized using statistics to guide human behavior.

Understanding Human Decision-Making: Advertising and Marketing

While life insurance companies were exploring the value of statistical analysis to predict and shape human behavior, manufacturers were facing a different set of problems. By the early 1900s, the Industrial Revolution was beginning to impact society in unanticipated ways. Increases in efficiency meant manufacturers could produce large amounts of goods to sell, but there seemed to be no way to increase the size of the market for these goods. Manufacturers turned to advertising and marketing to solve this problem. According to Stuart Ewen, "The functional goal of national advertising was the creation of desires and habits. In tune with the need for mass distribution that accompanied the development of mass production capabilities, advertising was trying to produce in readers personal needs,"[13] creating an ever-expanding marketplace. The problem became acute in the aftermath of World Wars I and II because of the massive manufacturing capabilities built to support the war effort, the "greatest production problem of any country in modern times."[14]

The problem was not getting people to buy; it was getting large numbers of people to buy consistently. As Vance Packard wrote in 1947, marketers believed the "disturbing difficulty was the apparent perversity and unpredictability of the prospective customers."[15] Solving the problem of retail inconsistency in order to build a new consumer-oriented market required something more subtle than the techniques used by life insurance companies. According to Packard, "What the [marketers] are looking for, of course, are the *whys* of our behavior, so that they can more effectively manipulate our habits and choices in their favor,"[16] by working with "the fabric of men's minds."[17] In 1951, McLuhan notes that this is the "first age in which many thousands of the best-trained individual minds have made it a

---

13. Ewen, *Captains of Consciousness*, 37.

14. Baime, *The Arsenal of Democracy*, 84.

15. Packard and Miller, *The Hidden Persuaders*, 37.

16. "Amazon.com Link," 32.

17. Packard and Miller, *The Hidden Persuaders*, 33.

full-time business to get inside the collective public mind. To get inside in order to manipulate, exploit, control is the object now."[18] By 2000, a system of designing retail shopping spaces was fully developed, including the arrangement of products in the space, to maximize economic productivity of shoppers by increasing spending and guiding shoppers towards purchasing items with higher profit margins.[19] Marketers believed sending anthropologists into stores to record consumer behavior and using focus groups could untangle the human decision-making process, allowing more consistent control over purchasing decisions.

Nudge

The desire of the advertising industry to create consistent consumers led to a good deal of research into the decision-making process itself. This result uncovered two systems, which Daniel Kahneman calls system 1 (also called "fast thinking") and system 2 (also called "slow thinking").[20] According to Kahneman, fast thinking "operates automatically and quickly, with little or no effort and no sense of voluntary control," while slow thinking "allocates attention to the effortful mental activities that demand it, including complex computations."[21] It is system 2 that is "associated with the subjective experience of agency, choice, and concentration."[22] Stephen Wendel writes, "Roughly half of our daily lives are spent executing habits and other intuitive behaviors," such as "gut instinct (blazingly fast evaluations of a situation based on our past experiences), or on simple rules of thumb (cognitive shortcuts or heuristics built into our mental machinery)."[23]

Sunstein and Thaler suggest that the concept of choice architecture, carefully arranging the conditions under which people make decisions, can be used to manage irrational decisions towards preferred results.[24] They call such changes in choice architecture a nudge.[25]

Nudging individuals by working with "the fabric of men's minds"[26] normalizes the use of "shortcuts" through the mental processes of humans

18. McLuhan, *The Essential McLuhan*, 21.

19. Underhill, *Why We Buy*, 4.

20. Kahneman, *Thinking, Fast and Slow*, 20.

21. Kahneman, *Thinking, Fast and Slow*, 20.

22. Kahneman, *Thinking, Fast and Slow*, 20.

23. Wendel, *Designing for Behavior Change*, loc. 739.

24. Thaler and Sunstein, *Nudge*, 7–8.

25. Thaler and Sunstein, *Nudge*, 9.

26. Packard and Miller, *The Hidden Persuaders*, 32.

to modify human behavior. Essentially, nudging attempts to guide system 2 decision-making in a predetermined direction. Nudging provides neurodigital media with one tool to manage the beliefs and actions of its users, and neurodigital media amplifies nudging by supplying broad surveillance and analysis based on that surveillance as well as an almost ubiquitous user experience through which to nudge users.

Markets of One

The development and widespread use of neurodigital media allowed the persuasive methods developed in advertising and retail to be applied using the personal levels of information provided through pervasive surveillance. Caroline Daniel and Maija Palmer tie the pervasive surveillance of neurodigital media to the effectiveness of advertising, writing, "The race to accumulate the most comprehensive database of individual information has become the new battleground for search engines as it will allow the industry to offer far more personalised advertisements."[27]

The result is a market of one—the ability to understand each individual person and attune advertising and marketing to their individual tastes and influences. Jason Jercinovic says,

> We are, all of us, sitting on a gold mine. For the last three decades or so, we have been collecting a treasure trove of digital information on everything from changing weather patterns to the spread of infectious diseases. We have digitized the history of the world's literature. We track and store the movements of automobiles, trains, planes and mobile phones. And we are privy to the raw, real-time sentiments of billions of people through social media. Individually, each of these digital resources has been immensely useful, applied to solving specific problems in dozens of industries. But collectively, when integrated, cross-referenced, and analyzed, this body of information represents the most powerful natural resource the world has ever known. And it is growing at a rate of 2.5 billion gigabytes every day. ... This capability holds profound implications for nearly every company in every industry. But for those of us in the marketing profession, it brings us ever closer to reaching a long-sought-after goal: markets of one.[28]

27. Daniel and Palmer, "Google's Goal."
28. Jercinovic, "Markets of One."

Jercinovic goes on to say that mass markets and their one-size-fits-all mentality will disintegrate, replaced by markets containing individualized goods and advertising.[29] Google founder Eric Schmidt argues for markets of one by emphasizing the use of data and analytics to make decisions and guide users through life. Schmidt says, "I actually think most people don't want Google to answer their questions, they want Google to tell them what they should be doing next."[30] So, he says, "The goal is to enable Google users to be able to ask the question such as 'What shall I do tomorrow?' and 'What job shall I take?'"[31]

According to Ryan Calo, "the digitization of commerce dramatically alters the capacity of firms to influence consumers at a personal level," empowering firms to "to discover and exploit the limits of each individual consumer's ability to pursue his or her own self-interest."[32] The goal is to "trigger irrationality or vulnerability in consumers," taking advantage of asymmetry between what users know about themselves and what organizations know about individual users.[33] The market of one represents extreme atomization; there will be no shared experiences, and indeed no reason to interact with people "in real life" (IRL) to achieve personal goals or happiness. Each market of one will be shaped by data to perfectly conform to the desires and fears of each person.

Nudging for Evil

Social recommender systems, discussed in some detail in chapter 5, are one prominent and common example of operators using choice architecture and nudges to guide consumers into behaving specific ways. These systems are joined by many other examples, however, many of which are more clearly designed to guide individuals toward more harmful actions.

Phishing is a form of social engineering where a threat actor sends an email that appears to originate from a trusted organization or individual, which causes the receiver to expose private information, such as bank account information or passwords. For instance, a phishing attack may consist of an apparent email from a bank asking the user to log into their account to validate their phone number or change their password. When the receiver follows the link, however, they find themselves at a web page that looks

29. Jercinovic, "Markets of One.".
30. Quoted in Jenkins, Jr., "Google and the Search for the Future."
31. Quoted in Daniel and Palmer, "Google's Goal."
32. Calo, "Digital Market Manipulation," 999.
33. Calo, "Digital Market Manipulation," 999.

correct, and may even be secured through encryption, but is a fake site set up by the threat actor to collect the requested information. Sophisticated phishing attacks will pass the user's information through to the real site so the user eventually ends up at the correct page on the real site corresponding to the fake site.

Michael Bossetta defines spear phishing as a phishing attack targeted at an individual (or set of individuals).[34] To form a spear phishing attack, threat actors mine publicly available information about a targeted information to personalize the attack, allowing the communications to be both more accurate and more urgent. According to Bossetta, attackers use analysis tools (much like advertisers) to identify targets (potential victims), assessing their access to private information and financial resources (wealth, position, and influence).[35] In some cases, threat actors also use information taken from previous data breaches, which can include information the target may believe is known only by organizations they already trust. Using information gleaned from these sources, the attacker can determine the best time of day to send a message, what organizations and people the target already trusts to bypass the target's rational thinking, tapping into faster decision processes to convince the target to take some action the attacker desires. Bossetta notes that attackers sometimes use elaborate identities created through social media to launch spear phishing attacks.[36] According to Symantec's 2018 "Internet Security Threat Report," "Spear phishing is the number one infection vector employed by 71 percent of organized groups in 2017."[37]

Spear phishing is often more effective at bypassing the slower thinking process in users, and the moral deliberations accompanying the slower process. Rather than using information gleaned from user information, however, spear phishing overcomes the user's ability to judge purely through deception—making the user believe they are taking a safe action, while actually causing the user to work against their own best interests.

Dark patterns are another form of nudge used to make users act against their own best interests. Ana Valjak writes, "Dark patterns are interfaces designed to deliberately deceive users in favor of a business," sneaky, money-driven elements that "are not user-centered experiences, but rather business-oriented tactics."[38] Rather than helping users toward a goal they

34. Bossetta, "Spear Phishing," 98.

35. Bossetta, "Spear Phishing," 100.

36. Bossetta, "Spear Phishing," 100–101.

37. "Internet Security Threat Report," 6.

38. Valjak, "Dark Patterns."

desire to the user's advantage, dark patterns guide the user towards an outcome that is advantageous to a business (or potentially a government entity) at the expense of the user. Valjak describes thirteen kinds of dark patterns, each of which disguises information, withholds information, takes advantage of a user's "not so great" decision making,[39] or nudges the user's decision in a way Valjak considers harmful to the user. Exploring some of these dark patterns is useful to provide context.

Hidden advertisements appear to be a search result or some form of navigation button (such as an item in a menu) instead of advertisements. According to Valjak, hiding advertisements in the areas of a page normally containing something other than advertising tricks the user into thinking they will receive some form of useful information they are looking for after clicking on the link; instead, they are taken to a site selling a product related to their query.[40] Jason Aten notes Google changed the layout of its search results in January of 2020 so advertisements are almost indistinguishable from pages requested by the user.[41] Fear of missing out (FOMO) uses various forms of pressure to force users to rely on faster decision-making processes, often resulting in unwise choices.[42] Mathur and colleagues describe several versions of FOMO,[43] including countdown timers, limited-time messages, "low stock" messages, and "high demand" messages, many of which are deceptive.[44]

Dark patterns overcome the user's moral reasoning by either withholding information, placing information in nonobvious locations, giving the user less-than-ideal options, or using visual tricks to convince the user they are choosing the best option when they are not. These user interface patterns can be used to override the user's moral reasoning more directly, often even if the user does engage their slow thinking systems, and makes a deliberate choice based on moral reasoning.

39. Thaler and Sunstein, *Nudge*, 12.

40. Valjak, "Dark Patterns."

41. Aten, "Google Is Making It Harder."

42. Valjak, "Dark Patterns."

43. The author of this book regularly gives technology webinars on a widely used publishing platform that uses FOMO dark patterns to increase sales. Users are presented with a deceptive "number of spots available" for each webinar. The author was informed, however, that the number shown is *not* the number of seats available, but rather is reduced "somewhat randomly" as the webinar date approaches to make users believe they will "miss out" if they do not sign up for the webinar.

44. Mathur et al., "Dark Patterns at Scale."

Habit

B. J. Fogg says using the word *break* to describe changing a habit is the wrong word, because "[i]t implies you exert sudden force and the habit goes away. A better verb would be 'untangle' because it sets the right expectations of how to get rid of such behaviors. It requires persistence."[45] Because of the persistence required, many people who try to break or untangle habits fail. Neurodigital media, however, is a nearly perfect platform for shaping human habits by providing a near real-time, data-focused feedback loop—what Stephen Wendel calls behaviorally effective design.[46] Habits fall between fast and slow thinking. According to Kahneman, a "crucial capability of System 2 is the adoption of 'task sets': it can program memory to obey an instruction that overrides habitual responses."[47] Given the right sequence, slow thinking can override fast thinking often enough to change it—untangling or forming a habit. Wendel outlines four steps to designing a product that can change user behavior: understand how humans decide, discover the right behaviors to change, design the product around the behavior to be changed, and refine the impact of the product on behavior.[48] The three latter steps—discovering, designing, and refining—are deeply enhanced (or in some cases made feasible) by neurodigital media. This section first shows that habit formation through neurodigital media is intentional, then how neurodigital media enables creating habit-forming user experiences, and finally provides real-world examples of habit-forming neurodigital media.

Habit Formation is Intentional

According to Benny Belvedere, Facebook intentionally "hooks" users into habitual use of the service because its "actual mission is maximal monetization by exploiting their users through monopolistic surveillance efficacy."[49] To do this, Belvedere says,

> [I]t needs you to spend as much time as possible on Facebook. To get you to do that, it needs to enable the sort of communal features that incentivize total immersion. If Facebook builds a place where individuals can gather to share memes and articles that reflect and reinforce group beliefs, it will increase the

45. Fogg, "Foreword," loc. 189.

46. Wendel, *Designing for Behavior Change*, loc. 255.

47. Kahneman, *Thinking, Fast and Slow*, 36.

48. Wendel, *Designing for Behavior Change*, loc. 309.

49. Belvedere, "Facebook Makes More If You're Addicted."

likelihood that people inside the group will seek to spend their time there.[50]

The Chief Executive Officer of Netflix says their service competes with a "broad range of things [users do] to relax and unwind," and "there are only a certain amount of hours which humans can tend to activities, and Netflix's goal is to occupy those moments—and deliver the utmost joy to the consumer during that opportunity."[51] Mohammed Bilal details how FlipKart, a shopping app, determined that using a common negative emotion, the fear of missing out, laid a basis for a habit-forming shopping experience to drive more consistent engagement between sales events, leading to more dependable purchasing behavior.[52] Nir Eyal writes, "Companies increasingly find that their economic value is a function of the strength of the habits they create."[53] Conor Henderson and colleagues argue that habitual purchases are an important element of consumer loyalty programs.[54] Causing consumers to habitually use a product, brand, or shopping experience is widely seen as crucial to corporate success.

Enabling Habit Formation through Neurodigital Media

Wendel says the discovery phase "clarifies what, specifically, the company wants to accomplish with the product, and for whom."[55] This phase includes understanding a market, a felt need within the market (in the case of habit formation a habit to untangle or create), and goals in terms of time and success (how fast, how long, and how the habit or the change in habit will be monetized). Discovery might be direct; a user might indicate a desire to lose weight or save money by signing up for a service or adjusting their preferences in an existing account. Discovery might also be indirect; profiles created using data gathered through surveillance, and analysis of that information through neurodigital media, might indicate a person is open to being nudged toward losing weight or saving money. For instance, sentiment analysis of the type described by Fabrício Benevenuto and other researchers could be used to find social media users who have a positive view of weight

50. Belvedere, "Facebook Makes More If You're Addicted."

51. Raphael, "Sleep Is Our Competition."

52. Bilal, "How to Design a Habit-Forming Shopping Experience."

53. Eyal, *Hooked*, 1.

54. Henderson, Beck, and Palmatier, "Review," 262.

55. Wendel, *Designing for Behavior Change*, loc. 309.

loss products, or a negative view of their own body.[56] If a widely recognized user (an influencer) is publicly promoting a diet plan or attempting to lose weight, the users they influence can be discovered using methods like those Aleksandr Farseev and his fellow researchers describe.[57] Wendel notes that assessing the motivation of potential users is very important to assessing the likelihood of achieving behavior change.[58] Level of motivation can be assessed using neurodigital media by examining a user's posts, actions, and connections, all of which are exposed through neurodigital media. Neurodigital media plays a critical role in discovering goals and individuals who will be open to pursuing those goals.

Design, according to Wendel, involves building a behavioral plan, or the "story about how the user will interact with the product," and a user experience.[59] The user experience is designed around what Wendel calls the create action funnel, which contains five mental events, "a *cue*, which starts an automatic, intuitive *reaction*, potentially bubbling up into a conscious *evaluation* of costs and benefits, the *ability* to act, and the right *timing* for action."[60] Neurodigital media makes it possible to control each of these five mental actions with some precision. An application installed on a personal device can cue the user by reminding them at a specific time of day, or when some event occurs, to take some action—taking care to give the cue when the user is able to act. This application can then nudge the reaction through choice architecture to guide the user along a predetermined path and modify the evaluation of costs and benefits by presenting some form of reward. Wendel provides a "straightforward recipe" for habit formation in a later section consisting of three elements: cue, routine, and reward.[61]

In order to build a habit, Wendel says the designer should identify the routine to be made into a habit, identify a reward that is meaningful and valuable for the user, and identify an unambiguous cue in the person's daily life or the application itself. Once these are identified, the designer should deploy the cue and facilitate the routine, then immediately reward the user.[62] Wendy Wood and David Neal, in a review of studies on consumer behavior, come to the same conclusion, saying, "Evidence from social cognitive

56. Benevenuto, Araújo, and Ribeiro, "Sentiment Analysis Methods," 11.

57. Farseev et al., "SoMin.Ai."

58. Wendel, *Designing for Behavior Change*, loc. 2551.

59. Wendel, *Designing for Behavior Change*, loc. 309. Wendell uses "user interface" here, but his description fits "user experience" more closely; see location 3091 for a discussion of the environment outside the application.

60. Wendel, *Designing for Behavior Change*, loc. 1400.

61. Wendel, *Designing for Behavior Change*, loc. 1789.

62. Wendel, *Designing for Behavior Change*, loc. 1789.

and from neuroscience research converge on the idea that habits are direct context-response associations in memory that develop with repetition."[63] All of these steps can be taken within the user experience of neurodigital media. Some form of reminder or notification can be used to cue the user, such as a sound, vibration, or even just a small red dot placed on the app's icon. Ezequiel Bruni, describing playing through a game even though "the game's story was an unrewarding, steaming pile of sadness," outlines rewards a user experience designer can offer, including connections to other people, popularity, recognition, competition, progress, achievement, exclusivity, discovery, and surprises.[64] Bruni says the emotional rewards have "enough emotional payoff to keep me around when I absolutely despised the central narrative. Imagine if a website (Facebook) could (Twitter) do (LinkedIn?) that (Amazon). Well they can, and they do."[65] Game designers use the same kinds of psychological rewards to make their product "sticky." Luyi Xu describes the use of artificial intelligence to make non-player characters more lifelike so players can emotionally connect with them more deeply, making playing more immersive.[66] Xu writes, "Games give players what they don't have in real life—the happiness of making friends, a feeling of superiority in their skills, and the feeling they are 'part of something.'"[67]

According to Wendel, "Once a version of the product is ready for field testing, the team starts to gather quantitative and qualitative data about user behavior to form an initial impact assessment of how the product is doing."[68] Designing and refining the product are iterative processes requiring rapid changes in user interface based on measurements made in near real time. These kinds of changes cannot be made in physical products quickly enough, but they can be made in software-based user interfaces. As Wendel points out, "If the behavior that the product is trying to change is part of the product itself, you're in luck. There are tools to help you gather the data."[69] Neurodigital media, being a form of digital media, allows designers to rapidly iterate through design options, testing each option directly for effectiveness. Wendel suggests using staggered rollouts as one method, where some users receive new features or changes in the user experience, while others do not; the effectiveness of the two versions of an application can indicate

63. Wood and Neal, "The Habitual Consumer," 580.

64. Bruni, "8 Ways to Emotionally Reward Your Users."

65. Bruni, "8 Ways to Emotionally Reward Your Users."

66. Xu, "Exploiting Psychology and Social Behavior for Game Stickiness," 52–53.

67. Xu, "Exploiting Psychology and Social Behavior for Game Stickiness," 52–53.

68. Wendel, *Designing for Behavior Change*, loc. 327.

69. Wendel, *Designing for Behavior Change*, loc. 4629.

whether or not the changes improved habit formation.[70] Arnold Vermeeren and company consider more than ninety different methods of evaluating user experience, many of which rely on quickly iterating through designs to test their effectiveness.[71]

Addiction

"Addiction is a state of compulsion, obsession, or preoccupation that enslaves a person's will or desire. Addiction sidetracks and eclipses the energy of our deepest, truest desire for love and goodness," according to Gerald May.[72] May additionally contends that addiction is "a self-defeating force that abuses our freedom and makes us do things we really do not want to do."[73] According to Anne Wilson Schaef, "An addiction is any process over which we are powerless. It takes control of us, causing us to do and think things that are inconsistent with our personal values and leading us to become progressively more compulsive and obsessive."[74] While addiction to systems built using neurodigital media is somewhat controversial, there is extensive evidence detailing compulsive behavior and withdrawal symptoms related to services such as Facebook.

Neurodigital Media and Addiction

Pamela Rutledge, writing for *Psychology Today*, says it "concerns me that, as a society, we are very cavalier tossing around the concept of 'addiction.' Addiction is a serious psychological diagnosis based on specific and seriously life-impairing criteria."[75] She observes that "addiction" is often used in relation to social media in order to "draw eyeballs to your copy because it targets people's fears."[76] Erin Brodwin argues against the possibility of addiction, saying, "Social media and smartphones are not ruining our brains, nor will either become the downfall of a generation."[77] She quotes a study by Daniel Kardefelt Winther to argue, "Teens are chiefly using digital

---

70. Wendel, *Designing for Behavior Change*, loc. 4853.

71. Vermeeren et al., "User Experience Evaluation Methods."

72. May, *Addiction and Grace*, 14.

73. May, *Addiction and Grace*, 3.

74. Schaef, *When Society Becomes an Addict*, 18.

75. Rutledge, "Social Media Addiction."

76. Rutledge, "Social Media Addiction.".

77. Brodwin, "There's No Solid Evidence."

communications to deepen and strengthen in-person relationships."[78] The study in question, however, does not entirely support Brodwin's line of argument. Winther does argue that recent neurological evidence challenges digital media's ability to "re-wire or hijack children's brains," but also that "considerable methodological limitations exist across the spectrum of research on the impact of digital technology on child well-being, including the majority of the studies on time use reviewed here, and those studies concerned with clinical or brain impacts."[79]

Two methodological problems are of importance. First, Adam Alter says, "Our understanding of addiction is too narrow" because "we tend to think of addiction as something inherent in certain people."[80] Instead, Alter insists, "Addiction is produced largely by environment and circumstance."[81] Vladan Starcevic and Elias Aboujaoude, in a review article, argue that the phrase *Internet addiction* is too broad, because "there seems to be little or no support for the construct of 'generalized pathological Internet use' (as opposed to 'specific pathological Internet use')."[82] Instead, they say users can become addicted to specific activities on the Internet.[83] For instance, Guangheng Dong and others report a strong correlation between the impairment of the anterior cingulate cortex, error processing, and addiction to Internet-based games,[84] a form of neurodigital media. Daria Kuss and Mark Griffiths hold that users can become addicted to social networking, which they differentiate from social media by the level of interactivity between users.[85] According to other researchers, "Social networking can be considered a way of being and relating," and the current generation find it "impossible to imagine life without being connected."[86] The authors say addiction to social networking may be related to the smartphone as a physical device, a fear of being without a mobile phone (nomophobia), or the fear of missing out.[87]

Second, according to Starcevic and Aboujaoude, the Internet as a medium may contribute to addictive behavior by making some activities more accessible, but it is unlikely that studies can be conducted to explore

78. Brodwin, "There's No Solid Evidence.".

79. Winther, "How Does the Time?"

80. Alter, *Irresistible*, 3.

81. Alter, *Irresistible*, 3.

82. Starcevic and Aboujaoude, "Internet Addiction," 11.

83. Starcevic and Aboujaoude, "Internet Addiction," 11.

84. Dong et al., "Impaired Error-Monitoring Function."

85. Kuss and Griffiths, "Social Networking Sites and Addiction," 2.

86. Kuss and Griffiths, "Social Networking Sites and Addiction," 5.

87. Kuss and Griffiths, "Social Networking Sites and Addiction," 8–9.

this possibility.[88] These observations indicate that neurodigital media may not be directly addictive like a physical substance, but is rather *seductive*, encouraging behavioral addiction to individual services or experiences. Alter says, "The environment and circumstance of the digital age are far more conducive to addiction than anything humans have experienced in our history."[89] He quotes Tristan Harris as saying, "The problem isn't that people lack willpower; it's that 'there are a thousand people on the other side of the screen whose job it is to break down the self-regulation you have.'"[90] Designs encouraging habitual use could easily lead to addictive behavior, especially given that addictive behavior is likely more profitable for the system operator.

Studies Suggesting Addictive Behavior

While defining addiction to services built on neurodigital media is difficult and controversial, it is possible to describe the effects of addiction, and then study the prevalence of these effects. One of the difficulties with studies of addiction is there is often no differentiation between the medium (the Internet or a smartphone, for instance) and the service (such as Facebook or some other neurodigital media). In this book, nomophobia, smartphone addiction, and self-interruption with smartphones and other digital devices are assumed to be related to the inability to access a specific set of services on the missing device, rather than the lack of a physical device itself. Researchers propose six indicators of addiction: salience, mood modification, tolerance, withdrawal symptoms, conflict, and relapse.[91] Studies provide examples of many of these symptoms in relation to neurodigital media use.

Salience is "[w]hen the online activity turns out to be the most important activity in the person's life and governs their thinking."[92] Salience can be shown by the engagement level of users with a service, primarily by considering the number of times a user accesses a service in a given time period, and how long they remain in the service. One study attempted to correlate smartphone addiction with smartphone interruption and productivity using a series of questions to evaluate addiction and self-reported interruption and productivity levels. The study concluded there was a positive correlation between the addiction level of the user and the number of times

88. Starcevic and Aboujaoude, "Internet Addiction," 10.

89. Alter, *Irresistible*, 3.

90. Alter, *Irresistible*, 3.

91. Vondrackova and Smahel, "Internet Addiction," 471.

92. Vondrackova and Smahel, "Internet Addiction," 471.

the user self-interrupted with a smartphone; the study also concluded there was a negative correlation between productivity and self-interruption with a smartphone.[93]

Another effect of salience is the offsetting of one activity by another; activities with higher salience will offset activities with lower salience. According to Joseph Turel and Michael Hennessy, time use studies show leisure Internet use (which is most likely some form of neurodigital media, such as social media or social gaming) has offset time spent on homework, time spent exercising, and time spent sleeping between 2012 and 2016.[94] Jean Twenge and company found a similar pattern in adolescent media use, with digital media partially offsetting television and almost completely offsetting printed media.[95] This result is not surprising, given Mary Meeker's report that average daily use of digital media increased from 2.7 hours in 2008 to 6.3 hours in 2016 in the United States.[96]

Mood modification is "experiencing a tranquilizing feeling of 'escape' or 'numbing' and a feeling of irritation when it is not possible to be online."[97] Researchers agree that separating users who are addicted to their smartphones causes a high level of stress; however, there is little understanding of the reason (or pathway) for this stress.[98] One study with young working professionals modeling different kinds of smartphone separation, concluded at least one reason for this elevated stress is a perception that being without a smartphone results in a social threat.[99] The authors do not explore *why* the lack of a smartphone is perceived as a social threat, but a likely candidate is the lack of a "retreat mechanism" on the part of the user.

Withdrawal symptoms consist of "[u]npleasant feelings, states, and/or physical effects (e.g., the shakes, moodiness, irritability) that come about when the specific online activity is discontinued or suddenly limited."[100] One study separated teenagers from their smartphones for three weeks. Subjects who were not highly engaged users showed no symptoms of withdrawal, while highly engaged users fell into two categories. One group self-reported finding the difficulty of living without their phone weakening

93. Duke and Montag, "Smartphone Addiction," 93.

94. Turel, "Potential 'Dark Sides,'" 25.

95. Twenge, Martin, and Spitzberg, "Trends."

96. Meeker, "Internet Trends 2019," 41.

97. Tams, Legoux, and Léger, "Smartphone Withdrawal," 1.

98. Tams, Legoux, and Léger, "Smartphone Withdrawal," 1.

99. Tams, Legoux, and Léger, "Smartphone Withdrawal," 6.

100. Rosen, Cheever, and Carrier, eds., *The Wiley Handbook of Psychology, Technology, and Society*, 472.

over the period of separation, while the second group found the difficulty of living without their phone increasing over time.[101]

Relapse occurs when a user decides to stop using the service, and later decides to begin using the service again. Eric Baumer and colleagues describe users citing surveillance, social pressure, and the ability to manage online perception as reasons why users relapse into social media use.[102] The authors used a combination of four methods to correlate the experience of withdrawal from Facebook with the tendency to relapse into use. The results of their study show that users reporting symptoms consistent with addiction were significantly more likely to relapse into using Facebook than those who said "nothing bad happened" when they stopped using the service.[103]

Addiction, Gambling, and Neurodigital Media

The number of people addicted to gambling is not well understood; surveys generally show about 1.2 percent of gamblers self-report as problem gamblers.[104] Mark Griffiths notes that technology innovation impacts problem gambling in several ways, including making gambling accessible via social networking sites, mobile gambling, and the increased use of behavioral tracking.[105] These uses of digital technology are not unique, however. Natasha Dow Schüll, in a deeply researched work on machine gambling, details the use of digital technologies to improve the time gamblers spend "on machine." Many of these techniques can be classified as neurodigital media.

Schüll's focus is on machine gambling, which "is associated with the greatest harm to gamblers"[106] because "individuals who regularly played video gambling devices became addicted three to four times more rapidly than other gamblers, even if they had regularly engaged in other forms of gambling in the past."[107] She attributes this increase in gambling to the entire experience of machine gambling (called user experience by more recent designers). The design of the gambling floor in a casino itself plays a role, shrinking space, focusing attention, and modulating experience so gamblers can enter into private playing worlds.[108] The experience of using

101. Adelhardt, Markus, and Eberle, "Teenagers' Reaction."
102. Baumer et al., "Missing Photos," 1.
103. Baumer et al., "Missing Photos," 7.
104. For instance, see "Levels of Problem Gambling in England."
105. Griffiths, "How Is Technology Innovation Impacting Gambling Addiction?"
106. Schüll, *Addiction by Design*, 14.
107. Schüll, *Addiction by Design*, 16.
108. Schüll, *Addiction by Design*, 40–50.

the machine, however, plays a much larger role in keeping the gambler "productive" for longer periods of time by accelerating play, extending the duration of play, and increasing the total amount spent.[109] Gamblers, or players as they are called throughout the gaming industry, seek the zone, according to Schüll, a state that is calm and relaxing rather than exciting.[110]

Virtual reel mapping is the first digital technology introduced to machine gambling, according to Schüll. Instead of using physical reels, digital gambling machines can play videos of spinning reels; this frees the game designer from mapping winning combinations to what the player sees, and allows the designer to precisely control the odds of winning. On a virtual reel, the player may only see three of a given symbol, but the virtual reel may contain many more. The odds the player perceives, then, are not the same as the odds the machine is using to calculate a win or loss.[111] Neurodigital machines can also be programmed to present many more "near misses" than actually occur by representing a winning combination just above or below the player's central row; this allows the designer to make it appear the player "almost won," and that the odds of winning are much higher than they really are. This prompts players to continue gambling because, as Schüll quotes one player saying, "You got close and you want to keep trying."[112] The user interface of digital gambling machines is designed so all the required controls are placed near to hand,[113] and can even include the capability to order food or other items so the player does not need to be disturbed, breaking their experience of the zone.[114]

In the 1990s, casinos began surveilling players to understand how to create environments and machines that would increase player productivity, either by increasing time on machine or increasing the rate at which players spend their money. This movement transformed gambling machines from stand-alone units to networked electronic surveillance devices to understand "how the technology might better coax continued play."[115] The result is that players can be tracked across many different venues, including shows, hotel shopping, and other areas, so casino managers can know more about them than they know about themselves.[116] These systems have been tied

109. Schüll, *Addiction by Design*, 52.
110. Schüll, *Addiction by Design*, 53.
111. Schüll, *Addiction by Design*, 86.
112. Schüll, *Addiction by Design*, 91.
113. Schüll, *Addiction by Design*, 53.
114. Schüll, *Addiction by Design*, 58.
115. Schüll, *Addiction by Design*, 144.
116. Schüll, *Addiction by Design*, 146.

to the financial records of individual players so a company can "calculate a player's 'predicted lifetime value,'" which is how much they are "likely to lose to the franchise over his or her lifetime," and which kinds of solicitations might bring a gambler with remaining value back to the casino.[117]

These digital computers calculating individual odds, combined with intimate surveillance to calculate precisely how often to reward or solicit each player, are an example of neurodigital media used to its fullest extent. The player's entire experience is controlled to produce the most economic value for the company running the system. The result is widespread habit formation in many and addiction in some smaller number of people. The players Schüll interviews describe sitting at a machine for fifteen hours straight,[118] having dreams of being chained to a machine until they had no more money to spend,[119] and rushing to the phone while asking an attendant to hold their machine so they can reschedule appointments and arrange things so they can continue gambling.[120] Emergency medical personnel showed Schüll videos of players continuing to gamble even as a person nearby collapses from a heart attack, and casino managers describing players continuing to gamble as the casino is flooded or smoke builds up from a fire in the building.[121]

Moral Dimensions of Machines Making Decisions

Nudges, habit formation, and addiction can all use neurodigital media to shortcut the human thinking process, either changing the fast thinking process or bypassing the slow thinking process. In the one case, modifying the fast thinking process can change a person's everyday habits and decisions; in the other, the slow thinking that might express free will is simply bypassed. When these shortcuts result in addiction, they are condemned (though often softly, as addictions can still be profitable to designers). Nudging and habit formation, however, are treated more ambiguously because they are perceived as neutral, able to be used either for good or evil.

---

117. Schüll, *Addiction by Design*, 153.
118. Schüll, *Addiction by Design*, 101.
119. Schüll, *Addiction by Design*, 179.
120. Schüll, *Addiction by Design*, 203.
121. Schüll, *Addiction by Design*, 31–32.

Arguments in Favor of Nudging and Habit Formation

The first argument often used by proponents is that using neurodigital media to nudge users and modify habits will create value for the entire community. AdExchanger, a company that sells advertising slots on web sites, surveyed influential people in advertising to discover what reasons they give consumers to justify behavioral tracking to make advertising more effective. According to one respondent, "[a]dvertising is the lubricant of the Internet," making the free flow of information many people count on possible.[122] Another respondent says targeted advertising lowers the cost of marketing, lowering the cost of goods.[123] Supporters of the gaming industry make similar arguments; according to Craig Davies, gambling supports 33,171 jobs and has a total impact of $6.3 billion yearly in the state of Pennsylvania.[124] The Canadian Gaming Association reports casino gambling, largely consisting of the kinds of neurodigital media-based machine gambling described by Schüll, adds more than $8 billion (Canadian) each year.[125] This kind of impact explains the kinds of "circular defense" Schüll describes in relation to the distortions of the way gambling machines work to influence gamers: "Presiding judges find that the practice is not fraud, because every regulatory laboratory approves it; regulatory laboratories like GLI claim that their job is not to scrutinize and uphold consumer protection laws, but only to test the features of machines that casino operators ask them to; manufacturers and casino operators claim that regulation is not up to them."[126]

A second, and more prevalent, argument of those supporting the nudge and habit forming widely implemented through neurodigital media is that these techniques can be used to improve human life. For instance, Samuel Salzer and Silja Voomla collected predictions for the future of behavioral science from leaders in the field, most of which include such sentiments as, "I see a future where behavioural science sits firmly in the world of social good. Humans run, and ruin, the world, and behavioral science helps us understand and drive changes in human behavior," and "many people around the world are showing a growing appetite for learning how they can nudge themselves towards better wealth, health and happiness."[127] Wendel states that the purpose of his work, *Designing for Behavior Change*,

122. AdExchanger, "If A Consumer Asked You."

123. AdExchanger, "If A Consumer Asked You."

124. Davies, "Gaming Industry Boosts."

125. "The National Economic Benefits of the Canadian Gaming Industry: Key Findings Report," 10.

126. Schüll, *Addiction by Design*, 90.

127. Salzer and Voomla, "Behavioral Design 2020 and Beyond."

is to help designers build products that help change behaviors their users already want to change—which is different from what he calls the "dark arts" of using these very same techniques to drive sales or change people's behavior without their knowledge of consent.[128] Thaler and Sunstein call the nudge "libertarian paternalism," writing, "The paternalistic aspect lies in the claim that it is legitimate for choice architects to try to influence people's behavior in order to make their lives longer, healthier, and better. In other words, we argue for self-conscious efforts, by institutions in the private sector and also by government, to steer people's choices in directions that will improve their lives. In our understanding, a policy is 'paternalistic' if it tries to influence choices in a way that will make choosers better off, as judged by themselves."[129]

Counterarguments

Contrary to the arguments of the supporters of these methods, they will or should only be used for the good of users. Calo argues that, once invented and perfected, such methods will be used for digital market manipulation by "nudging for profit."[130] Calo says the rise of digital technologies has created a "mediated consumer" where companies monitor consumer behavior and control every interaction with the consumer, including when to approach consumers (rather than waiting on consumer to approach the company).[131] The ability to completely control consumer's interaction with companies, combined with surveillance, allows companies to stop theorizing about why certain causes result in individual behaviors and focus on direct testing to determine how to change behavior—what Chris Anderson calls "the end of theory."[132] The trouble, according to Calo, is "when firms start looking at the consumer behavior data set to identify consumer vulnerabilities,"[133] noting there is no "difference between placing impulse items by the counter and texting an offer to consumers when they are at their most impulsive."[134] For instance, Alain Samson describes how he discovered the use of surveillance

---

128. Wendel, *Designing for Behavior Change*, loc. 454.

129. Thaler and Sunstein, *Nudge*, 8.

130. Calo, "Digital Market Manipulation," 1001.

131. Calo, "Digital Market Manipulation," 1004.

132. Anderson, "The End of Theory."

133. Calo, "Digital Market Manipulation," 1010.

134. Calo, "Digital Market Manipulation," 998.

to trigger a personal follow-up that included several kinds of psychological pressure in response to his search for a product on a shopping site.[135]

A second ethical problem with nudging and intentional habit formation through neurodigital media is that it can pit the moral decision-making of the individual against the decisions of the larger culture, giving organizations representing the culture new tools to enforce societal norms. A simple case is the imposition of daylight saving time by many countries. According to David Berreby, many people believe moving clocks forward and backwards once a year (to and from daylight savings time) conserves enough energy to justify the disruption to the lives of individuals, but others do not. In this kind of situation, Berreby says, "Even the guy who invented nudge once applied the label to a policy that is non-consensual and not obviously desirable to the people it affects."[136] While Berreby's example is not directly related to neurodigital media, it does show the blurred line between Sunstein and Thaler's nudge and what Berreby calls a "shove."

Even if these technologies were only used for good, or with the best intentions, the results can still be damaging to individual users. According to Adam Alter, the problem with the set of tools called neurodigital media here is that "we've made the problem worse by focusing on the benefits of goal-setting without considering its drawbacks."[137] Linda Anticoli and Marco Basaldella describe three scenarios resulting from the quest for fitness, which they consider to be limitless, instead of health, which they consider objective and measurable.[138] Their third scenario, *Charlie's Problem*, illustrates the power of neurodigital media to fitness. In this scenario, a child suffers from kidney failure, but is placed lower on the list of priorities for a transplant because surveillance of his activities at home and school are analyzed, resulting in his being labeled "lazy." The only way for this child to improve his score, moving up the transplant list, is to exercise more and eat better.[139] While the situation Anticoli and Basaldella describe is fictional, it is quite plausible; insurance companies and medical practitioners have a strong interest in insuring available care is apportioned to provide optimal results and ensure that patients are motivated to improve their health (for their own good). In this case, however, the quest for fitness turns into a race against others who are competing for the same care.

---

135. Samson, "Big Data Is Nudging You."
136. Berreby, "Is It a Nudge or a Shove?"
137. Alter, *Irresistible*, 5.
138. Anticoli and Basaldella, "Shut up and Run."
139. Anticoli and Basaldella, "Shut up and Run," 1555–56.

Setting the goal to "better," with no standard by which "better" can be measured other than "what everyone else is doing," initiates a race to the top (or bottom), forcing individuals to conform and outperform at ever-higher levels. This race to the top echoes the view of G. S. Stent on the nature of progress, who says,

> That is, the better world is one in which man has greater power over external events, one in which he has gained a greater dominion over nature, one in which he is economically more secure. . . . This definition does not, therefore, encompass such wholly internal aspects of the human condition as happiness. Hence, it is a totally amoral view of progress, under which nuclear ballistic missiles definitely represent progress over gunpowder cannonballs, which in turn represent progress over bows and arrows.[140]

Stent ties this view directly back to a nihilistic will to power.[141] Moral decision-making, such as trading health off against other potential goods, is overcome by a quest for an unachievable goal. Human freedom is replaced by a quest regulated through neurodigital media in each area of life this technology invades.

Underlying these problems in applying neurodigital media to nudge or modify habit formation is a shared issue with the definition of *good*. When neurodigital media is used to overrule individual moral choices for financial gain or to create a more desirable society, good is taken to mean the improvement of the group overtaking the free decisions of individuals within that group. When neurodigital media is used to allow the user to improve themselves, good is taken to mean adjusting or "shortcutting" free decisions in the future in order to reach some proposed goal. While overriding moral freedom is not necessarily unethical in either case, it can become unethical if the good is not properly defined.

What is often missed is that it's not the machines that are controlling other people, but the people who design, develop, and control these neurodigital media systems. Lewis's prescient warning was true: "Each new power won by man is a power over man as well. Each advance leaves him weaker as well as stronger," with the result that man has conquered himself.[142] These technologies are just another set of tools one group of humans can use to control another group of humans.

---

140. Stent, *The Coming of the Golden Age*, 90.
141. Stent, *The Coming of the Golden Age*, 90.
142. Lewis, *The Abolition of Man*, 55.

9

Community

THERE IS LITTLE QUESTION that community is dying in the modern world. The "little platoons" (Burke) that hold a culture together are themselves being pulled apart by seemingly irresistible forces. From a Judeo-Christian viewpoint, this breakdown in community is not only bad at a societal level, it also directly harms each person by denying them the relationships necessary to flourish as a human.

While the widespread deployment of neurodigital media cannot be solely blamed for these cultural changes, it is helping these harmful changes along. A quick review of the arguments raised thus far will help bring the impact of neurodigital media on communities into sharper focus. The designers, developers, and operators of neurodigital media systems work within a culture holding a naturalistic view of the person. In this view, individuals are the simple product of human evolution, and have no dignity beyond the economic or cultural value assigned to them. This naturalistic view gives permission to operators of neurodigital media systems to shape users towards specific economic and cultural (normally progressive) ends. Neurodigital media, then, encourages humans to treat other humans as objects, and then provides the theory and tools to manipulate those "objects" (other humans). Whether intentional or not, these techniques break down the communities so important to society and the flourishing of the person.

How does neurodigital media destroy community? First, it encourages the atomization of society into communities of one—creating users who appear to be hyperconnected, but really stand alone against the world. This occurs because of filter bubbles and chilling speech (or the spiral of silence). Second, it hollows out relationships by encouraging virtual friendships—which do not fulfill the same spaces in our lives as real ("in real life," or IRL) relationships—and by commodifying relationships.

Filter Bubbles

Using a service that only plays the kinds of music you like sounds like a great idea—there is a good chance you will like most of the songs played, and you might discover some new music you like. Or imagine a dating site (if you are in that phase of life!) that always finds the perfect match. Rather than spending hours looking through profiles, you can just wait for the algorithm to offer up the next person you should date.

These might sound like wonderful ideas. Not only would these things save you a lot of time, they would also help you explore a world of options without being constantly disappointed in what you might find. On the other hand, these things can limit your choices in unhealthy ways. Boredom is only the beginning of the stunting of thought and imagination that can happen when you only listen to things you like.

In a way, each of these examples is a filter bubble; the system filters out things determined by data analytics you will not be interested in. Filter bubbles "create a unique universe of information for each of us . . . which fundamentally alters the way we encounter ideas and information."[1] If the point of a neurodigital media system is to get your attention, there is good reason to only show you things you will like because you are more likely to emotionally engage in those things.

Filter bubbles operate in two distinct ways—organically by users choosing what to share through the social network, and through the filtering of information based on what operators learn about user, as shown below.

1. Pariser, *The Filter Bubble*, introduction.

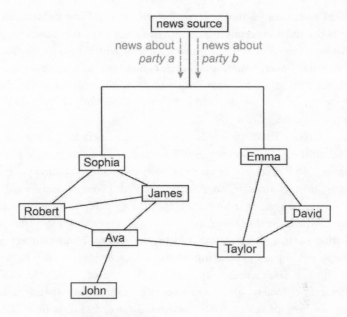

This figure illustrates the connections between users and a news source in a social media network. The system presents stories to each user on some sort of timeline. One should assume the news source produces two kinds of items, one of which favors *party a*, while the other favors *party b*. Sophia, who favors *party a*, and Emma, who favors *party b*, are each more likely to share items with which they agree to their followers, resulting in David and Robert receiving items favoring only one of the two parties. Ava may also share items favoring *party a* with Taylor, but this is not likely because of the reaction she is likely to get from Taylor by doing so. As explained in a later section, social media tends to chill speech, causing users to seek out an audience who agrees with them.

Even if Ava shares content with Taylor, she cannot be confident it will appear on his timeline because operators filter information to maximize engagement and promote the operator's vision of reality. Sophia and Emma, who might be called influencers in this illustration because of their position between the news source and other users, are important in forming the filter bubble. Because these two users have limited amounts of time to read and share items, they are likely only to share some of the items that appear on their timeline. William Brady and his fellow researchers found that "moral and emotional content are prioritized in early visual attention more than neutral content, and that such attentional capture is associated" with increased resharing.[2] Kalev Leetaru likewise notes, "The unfortunate fact that it is the most entertaining comment rather than the most enlightening that tends to go viral and reward its author with fame in today's world, [meaning that] we are teaching an entire generation to focus on information in isolation."[3] Further, because Emma and Sophia both have clear preferences for one party over the other, they will likely share only information about the party they favor.

The result of this organic filtering by users to increase their influence is only the most emotionally engaging information favoring *party a* is shared by Sophia, and Robert will not receive any information favoring *party b* (or even any "boring" material supporting *party a*). Likewise, Emma will only share the most engaging information supporting *party b*'s position with her followers, which means David and Taylor will not receive any information supporting *party a*. The location and preferences of the influencers in the social network form a natural information filter, creating a filter bubble for their followers. The mere operation of a service based on neurodigital media containing social media networks creates communities that do not have a complete view of reality. These communities exist inside filter bubbles, where they will only receive information that supports their preexisting ideas or beliefs.

Organic filtering, however, is less effective at creating filter bubbles than content filtering on the part of operators. The first reason operators filter information inserted in users' timelines is to increase engagement. Claudio Lombardi says, "Facebook's News Feed algorithm utilizes hundreds of variables to predict what is relevant for each user, based on 'relevance scores' that predict what a user will 'like,' comment, share, hide, click, or mark as spam. Advertisements as well as user-generated content all receive relevance scores."[4] Knowing each user only has a limited amount of time and attention, operators use analytics on the information they collect from each user (see the earlier discussion of pervasive surveillance) to discover the kinds of items each user will react to and adjust the information presented on the user's timeline to present only these kinds of items. If the operator knows Sophia favors *party a*, it may place a filter on Sophia's timeline, so she only receives items favoring *party a*. In this case, Sophia would not have the opportunity to share information she disagrees with because those items never appear in her timeline.

2. Brady, Gantman, and Van Bavel, "Attentional Capture," 3.

3. Leetaru, "How Social Media Is Teaching Us."

4. Lombardi, "The Illusion of a 'Marketplace of Ideas' and the Right to Truth," 3.

Returning to the situation where Ava shares information about *party a* with Taylor—algorithms the operator creates to determine which items to place on Taylor's timeline will probably note he does not agree with *party a*, so placing items in his timeline supporting *party a* will likely upset him, which will reduce his engagement on the platform. Unless the operator is intentionally trying to upset Taylor, he is not likely to see any items relating to *party a* that Ava might happen to share. Pariser describes connecting with several users of different political persuasions simply so he can understand what they believe, only to find Facebook would not place items these connections shared into Pariser's timeline—Facebook knew the author's political persuasion and simply filtered all other views out of the information presented.[5]

Finally, operators may generally filter information to remove items it deems unattractive or inappropriate or even items the operator opposes on social or political grounds. Sam Biddle and his colleagues describe an internal memo leaked to the public showing that TikTok, a social media service based on sharing short videos,[6] de-emphasizes (or does not place into user's timelines) videos containing people who are poor or "not attractive."[7] Most social media networks also have community guidelines to control the content users are allowed to post. Facebook's community standards state, "We remove content that glorifies violence or celebrates the suffering or humiliation of others because it may create an environment that discourages participation."[8] A similar rule on Twitter has generated controversy because, according to writers such as Seth Frantzman, it is unequally applied depending on whether the operator politically supports the user.[9]

Social media services can also "shadowban" content the operator does not want to be widely viewed but that does not directly violate community guidelines. Shadowbanning is described in a patent as a system where "the social networking system may receive a list of proscribed content and block comments containing the proscribed content by reducing the distribution of those comments to other viewing users. However, the social networking system may display the blocked content to the commenting user such that the commenting user is not made aware that his or her comment was blocked."[10] While shadowbanned items appear to be just like any other post

5. Pariser, *The Filter Bubble*, introduction.

6. "About TikTok."

7. Biddle, Ribeiro, and Dias, "Invisible Censorship."

8. Facebook, "Community Standards."

9. Frantzman, "Twitter Censors Trump."

10. Kanter, Singh, and Muriello, moderating content in an online forum.

to the sharing user, they are not considered for inclusion in other users' timelines because the operator does not want to allow the point of view represented to be widely distributed.

Researchers have widely documented the existence of filter bubbles. Robert Epstein and Ronald Robertson detail a series of experiments showing the suggestions given to users when typing into a search engine are biased by the operator, potentially impacting tens of millions of voters and directly changing the results of elections; they call this the "Search Engine Manipulation Effect."[11] Muhammad Ali and his fellow researchers document a filter bubble in Facebook's political advertising, saying, "Our findings point to advertising platforms' potential role in political polarization and creating informational filter bubbles."[12] John Wihby and company study the news media ecosystem in the United States, showing how reporters' and editors' reliance on Twitter shapes how and what reporters cover.[13] Shannon McGregor and Logan Molyneux also document journalists' reliance on Twitter,[14] and Lenny Bernstein and coauthors note, "Editors often assign stories based on what is trending and what competitors or sources are saying on Twitter," citing an internal memo to the staff of a large news organization.[15] The search engine DuckDuckGo performed a study showing the effect of the filter bubble on search results through Google, saying "The filter bubble is particularly pernicious when searching for political topics."[16]

Filter bubbles modify users' view of reality, creating intended and unintended repercussions. In 2018, a short video clip, taken from a much longer sequence, of a young man facing an older Native American in the United States went viral, driving millions of social media posts and thousands of stories in the news media. Those sharing and reporting on the original story found a single incident that seemed to support their worldview and amplified using social media to shame those with whom they disagreed into accepting their vision of reality. The filter bubble became a weapon of culture, a self-reinforcing feedback loop supporting a worldview. Careful investigators, however, found that the initial narrative, driven by a short video clip and a single image taken from that clip, to be a false view of reality. Ian Bogost shows that the original video when placed in a broader context of other videos of the same event, reverses the narrative provided

11. Epstein and Robertson, "The Search Engine Manipulation Effect."

12. Ali et al., "Ad Delivery Algorithms," 1.

13. Wihbey, Joseph, and Lazer, "The Social Silos of Journalism?"

14. McGregor and Molyneux, "Twitter's Influence on News Judgment."

15. "Recommendations for Social Media Use on the National Desk."

16. DuckDuckGo, "Measuring the Filter Bubble."

by major news organizations amplifying—and being amplified by—content being shared and reshared on Twitter.[17]

A little noticed effect of filter bubbles is that they cut people and cultures off from the past as well as present reality. The sheer amount of information each person is presented within each moment—the stream of videos, photographs, likes, memes, connections, and trivial bits of "news"— overwhelm the user. The amount of information draws users into the world of neurodigital media through constant reminders of what they are "missing," overwhelming the users' ability to pay attention, relate, and internalize. There simply is no room for the past in this stream, other than as a source of memes—as Jonathan Haidt and Tobias Rose-Stockwell say, "Even if social media could be cured of its outrage-enhancing effects, it would still raise problems for the stability of democracy. One such problem is the degree to which the ideas and conflicts of the present moment dominate and displace older ideas and the lessons of the past."[18]

Filter bubbles reduce community formation by altering the reality of individuals, so that they are only aware of a narrower range of ideas. Lombardi says, "Every internet user occupies an online environment tailored for him or her, polarized among different 'communities of interest,' each segregated in such a way that network effects are amplified. The result is that internet users' environments are populated with content primarily from their own community."[19] It is difficult to see how two users completely embedded in two such communities could form relationships by discussing issues or ideas, as they are, according to Vaidhyanathan, "incapable of engaging with each other upon a shared body of accepted truths."[20] Peter Pomerantsev claims that in cases where the very nature of reality is at stake in the discussion, "the side that is less constrained by the truth may be more likely to win."[21]

Chilling Speech

No one likes to disagree with a room full of people, especially if those people are actively engaged in arguing a deeply held belief. Elisabeth Noelle-Neumann contends, "The climate of opinion depends on who talks and who

17. Bogost, "Stop Trusting Viral Videos."
18. Haidt and Rose-Stockwell, "The Dark Psychology of Social Networks."
19. Lombardi, "The Illusion of a 'Marketplace of Ideas' and the Right to Truth," 3.
20. Vaidhyanathan, *Antisocial Media*, 10.
21. Pomerantsev, "Beyond Propaganda."

keeps quiet,"[22] and describes a process where those who disagree with a popular stance "stop talking," giving the impression the popular stance is gaining in popularity, in turn causing more individuals to stop speaking up to support opposing views.[23] She calls this the "spiral of silence," writing, "The fear of isolation seems to be the force that sets the spiral of silence in motion."[24]

This spiral of silence is not just theoretical. Andrew Hayes and other researchers performed social experiments showing some portion of the population does, in fact, possess a strong "willingness to self-censor" related to the hostility of the environment in which they find themselves.[25] Jonathan Penney studied the behavior of users to show even that the perception of government surveillance—even when users were not searching for ostensibly illegal material—had a measurable effect on their willingness to seek out information contrary to what they perceived would be considered acceptable by those who were watching.[26] In a study reported by Emily Ekins, 62 percent of American stated they held political opinions they were afraid to express because they feared losing their jobs or other social consequences.[27]

This spiral of silence applies to the online, digital world as well as the physical one. Social media platforms often exhibit extreme hostility to minority or unpopular opinions by canceling the people who hold them. While cancel culture is difficult to define, the effects are real. Benjamin Gill describes a large church whose contract with a local city to hold services in public schools were canceled because of objections to the pastor's reaction to material posted on Twitter.[28] Brooke Steinberg describes Disney closing long-popular rides because they are perceived as racist, with the action driven by social media users.[29] According to Madeline Osburn, a popular food writer, Alison Roman, lost her job at the *New York Times* for making comments during an interview, enraging an online group.[30]

When the consequences become losing a job, the pressure to stay within the bounds considered acceptable by even a small minority can cause

22. Noelle-Neumann, *The Spiral of Silence*, 4.

23. Noelle-Neumann, *The Spiral of Silence*, 5.

24. Noelle-Neumann, *The Spiral of Silence*, 6.

25. Hayes, Glynn, and Shanahan, "Validating the Willingness to Self-Censor Scale," 444.

26. Penney, "Chilling Effects."

27. Ekins, "Poll."

28. Gill, "Megachurch Given the Boot by City."

29. Steinberg, "Disney Fans Say Splash Mountain Needs Update."

30. Osburn, "How Chrissy Teigen."

self-censorship, silencing the thoughts, engagement, and voices needed to form communities. Jonathan Kay describes the impact on individuals:

> When they watch their likes and retweets and such, they are like investors of old watching a stock ticker. To lose status on social media is, for this cohort, a form of bankruptcy. That's why they run scared from controversial opinions, pressure colleagues within their organization to respect the Twitter-enforced party line, and join the pack against excommunicated heretics. In this environment, each of them secretly knows that they could be next. Nor can all of this be treated as a compartmentalized online phenomenon that has little bearing on the world outside of Stupid Twitter-land: The pathologies that now have become normalized on social media are beginning to metastasize to the real world.[31]

James Riley, whose social media accounts were blocked because he posted content other users complained about, writes,

> We "kid" each other about being in jail. Stop it. It's not funny. There are people I can't contact right now. There are ads I can't place for my business. There's an annoying "hate speech" pop-up I have to read whenever I mistakenly "like" or attempt to "share" something. If you were at a cocktail party, and the host used three goons to put gorilla tape over the mouth of a guest, I suppose a twisted type might laugh, but it's different when you are the one being gagged. When you are put in social media jail, it's very much like becoming an untouchable, an "outside the camp and unclean" person. You can watch. You can be lectured. But you can't respond.[32]

Kay says, "The idea that a whole career can fall victim to a single social-media message sent in a moment of anger or frustration—or even a bad joke—has produced an atmosphere of real terror that is compromising the art and intellect of Canada's most creative minds."[33]

Governments have taken note of the power of these technologies to distort and control the community. Gary King and his cohort performed an in-depth study of the Chinese government's censorship using interviews and informants to uncover its extent and purposes. While many observers hold that Chinese censorship is aimed at preventing criticism of the government, King's research found "that, contrary to much research and

31. Kay, "The Tyranny of Twitter."
32. Riley, "The Frightening Power of Social Media Bans."
33. Kay, "The Tyranny of Twitter."

commentary, the purpose of the censorship program is not to suppress criticism of the state or the Communist Party. Indeed, despite widespread censorship of social media, we find that when the Chinese people write scathing criticisms of their government and its leaders, the probability that their post will be censored does not increase."[34] Through this research, they found "the purpose of the censorship program is to reduce the probability of collective action by clipping social ties whenever any collective movements are in evidence or expected."[35]

Hollowing Out Relationships

Nellie Bowles, writing in *The New York Times*, says that when computers and smartphones were first introduced, it was a status symbol reserved for the wealthy to own and carry. Over time, *not* carrying any sort of device has become a status symbol, and human contact itself has become a "luxury good."[36] Eric Andrew-Gee quotes a former vice president of user growth at Facebook as saying, "The short-term, dopamine-driven feedback loops that we have created are destroying how society works."[37] Vaidhyanathan says that while the creators of social media originally intended to design systems that would connect the world, "the idealistic vision of people sharing more information with ever more people has not improved nations or global culture, enhanced mutual understanding, or strengthened democratic movements."[38] The result, according to Nolen Gertz, is "social media is the flash mob," carrying "shame campaigns"[39] and what Turkle calls "post-familial families," whose members are "alone together.'[40] Turkle says, "The Japanese take as a given that cell phones, texting, instant messaging, e-mail, and online gaming have created social isolation. They see people turning away from family to focus attention on their screens. People do not meet face to face; they do not join organizations."[41]

There are likely many causes for the hollowing of modern relationships, but neurodigital media set within naturalistic anthropology, which holds that relationships have no value beyond the instrumental, clearly

34. King, Pan, and Roberts, "How Censorship in China," 1.
35. King, Pan, and Roberts, "How Censorship in China."
36. Bowles, "Human Contact Is Now a Luxury Good." .
37. Andrew-Gee, "Your Smartphone Is Making You Stupid."
38. Vaidhyanathan, *Antisocial Media*, 5.
39. Gertz, *Nihilism and Technology*, loc. 3239.
40. Turkle, *Alone Together*, 279.
41. Turkle, *Alone Together*, 145.

contributes to the phenomenon. The first subsection below considers research exploring the relationship between neurodigital media and the loss of relationships. The second considers the argument that the virtual friendships formed on services built using neurodigital media are less than full friendships. The third discusses the commodification of relationships.

Statistical Evidence of Relational Harm

Because studying and quantifying the full damage caused by widespread use of systems built on neurodigital media is difficult, most studies focus on one or two components of the system. Valentina Rotondi and company performed an extensive study that measured life satisfaction, time spent with friends, and smartphone use. The researchers discovered that "time spent with friends is worth less, in terms of life satisfaction, for individuals who use the smartphone," supporting "the hypothesis that the smartphone negatively affects the quality of face-to-face social interactions."[42] They further found that although smartphone use, in general, contributed to an increase in life satisfaction, this positive effect was wholly offset by the decrease in time spent with friends, resulting in overall lower life satisfaction for more consistent users of smartphones.[43]

Ron Hammond and his colleagues write,

> The results of a multivariate analysis, based on 509 respondents, indicated that using Facebook is negatively related to the satisfaction with and quality of intimate relationships. First, the more often respondents updated their Facebook, the less satisfied they were with their spouses or partners. Second, the more years respondents used Facebook, the more likely it was for them and their partners or spouses to see defects in each other and to be defensive with each other. Third, the more "friends" respondents added to their Facebook, the more likely it was that they and their partners or spouses withdrew from each other. Fourth, those who spent more time with friends offline were more satisfied with their relationships with their partners/spouses, and they were also less likely to withdraw from their partners/spouses.[44]

Brian Primrack and his cohort conducted a study to determine the relationship between social media use and perceived social isolation. They

42. Rotondi, Stanca, and Tomasuolo, "Connecting Alone," 18.
43. Rotondi, Stanca, and Tomasuolo, "Connecting Alone," 21.
44. Hammond and Chou, "Using Facebook," 41.

found that "among a nationally representative cohort of individuals aged 19–32 years, there were robust linear associations between increased [social media use] and increased [perceived isolation], even after adjusting for a comprehensive set of covariates."[45]

Divorce is a narrow case of social media interference in relationships that has been more intensely studied. Sebastián Valenzuela and coauthors discovered that the penetration of Facebook into a culture or community is a "positive predictor of divorce rates."[46] According to their research, "a 20% annual increase in Facebook penetration rates is associated with an average 4.32% growth in divorce rates."[47] Jesse Fox argues that one reason for the correlation between marital satisfaction and Facebook use is that partners surveil one another using the service, resulting in negative feelings of mistrust and the potential for jealousy between partners.[48] Rianne Farrugia studied the impact of Facebook specifically with long-term relationship health and found "Facebook usage correlates with jealousy," because their spouse can remain in contact with partners from older relationships.[49] Russell Clayton and his fellow researchers write, "High levels of Facebook use may also serve as an indirect temptation for physical and/or emotional cheating" when a partner adds an ex-partner or spouse as a friend or begins to communicate with an ex-partner through private chat or messaging.[50] While not directly related to social media networks, Rose McDermott found divorce can be "contagious" within a social network, "like a rumor."[51]

While divorce is the most extreme expression of marital unhappiness, couples often remain married although they are unhappy with the relationship. Davies and company found that the wife's use of *Facebook* is a good predictor of marital unhappiness in their large-scale study.[52] They also found the wife's perceived social isolation was negatively correlated with the husband's investment in the marriage, likely leading to further marital stress.[53]

45. Primack et al., "Social Media Use and Perceived Social Isolation," 5.

46. Valenzuela, Halpern, and Katz, "Social Network Sites," 97.

47. Valenzuela, Halpern, and Katz, "Social Network Sites," 36.

48. Fox, "The Dark Side of Social Networking Sites," 78.

49. Farrugia, "Facebook and Relationships," 32.

50. Clayton, Nagurney, and Smith, "Cheating, Breakup, and Divorce," 719.

51. McDermott, Fowler, and Christakis, "Breaking Up Is Hard to Do," 17.

52. Davies et al., "Habitual, Unregulated Media Use and Marital Satisfaction," 77.

53. Davies et al., "Habitual, Unregulated Media Use and Marital Satisfaction," 77.

Virtual Friendships

Evidence that systems built on neurodigital media negatively impact the formation and strength of relationships must connect to relational harm through a theory of why or how it causes this harm. Dean Cocking and Steve Matthews, in a seminal paper on the topic of virtual relationships, say friendship has three essential components: "intimate relationships in which there is deep mutual affection, a disposition to assist in the welfare of the other, and a continuing desire to engage with the other in shared activities."[54] They argue that interpretation is central to all three of these aspects of friendship, which they define as the "interpretation of a friend's character and the ways in which we are consequently moved to relate to one another."[55] Further, "it is upon the interpretations of character between close friends that mutual affection, the desires for shared experiences, and the disposition to benefit and promote the interests of one's friend are expressed."[56]

An example of interpretation, in line with Cocking and Matthew's argument, might be seeing a friend become excited about a new camera and inferring the person enjoys photography. As a result, the friend who wants to grow closer will attempt to learn about good places and times to capture beautiful photographs and arrange to spend time with that person in those places and at those times. Enabling shared experiences with others is a way to create quality time and deepen the friendship. These shared experiences are enabled through a disclosure between the two people, an expression of excitement, and the ability to join in a shared physical activity, which results in a deepening relationship.

According to Cocking and Matthews, however, this kind of self-disclosure is simply not possible in the virtual world. First, Cocking and Matthews argue that these self-disclosures are not possible because each person can carefully control the image they present in virtual worlds. As described in chapter 3, neurodigital media encourages self-flattening through carefully managed self-presentation and other-flattening due to the presentation of a subset of the whole person. Both forms of flattening work against the kinds of self-disclosure Cocking and Matthews argue are crucial to forming real friendships. As Micah Meadowcroft says, social media disrupts this self-disclosure "so that we see not people, but discrete statements, abstracted ideologies, and caricatures. A tweet, comment, post, picture, fave, like, or

54. Cocking and Matthews, "Unreal Friends," 226.
55. Cocking and Matthews, "Unreal Friends," 227.
56. Cocking and Matthews, "Unreal Friends," 227.

share is served up as an act of the mind so distinct from anything else that there is no natural, no tacit, comprehensive awareness of the mind behind them."[57] According to Meadowcroft, users do not interact with one another on social media; instead, they interact with datums.[58]

This lack of self-disclosure involves the timing of interaction as well as the content. Cocking and Matthews say,

> The nature of my responses to others in the virtual world also diverges from the way we ordinarily respond to our friends. First, it is up to me when I respond to their contacts in ways that are unavailable in the non-virtual context; there will be no uncomfortable pause, no faux pas, when I hesitate briefly to construct a more carefully honed response. Second, my responses can be made without being interrupted, talked over, or qualified in other ways involving my being subject to the thoughts of others. And, of course, I can choose whether or not I will respond at all.[59]

This ability to warp the timing of interactions is, according to R. C. Victorina, a boon to productivity because "it takes the average person 25 minutes to return to their original task after being interrupted," and each synchronous communication event is an interruption.[60] This distortion in time, however, disrupts what Meadowcraft calls "*Kairos,* the opportune time, the time for actions in space,"[61] and "heart of social media's distortive power is . . . a separation of life's unity of time and space."[62]

Cocking and Matthew's concept of interpretation also includes a link between action and consequence; it is because one person cares for another that they take actions towards the consequence of a deepening friendship. The friend who finds opportunities for photography to please another is creating real-world consequences because of their shared connection with that person. Contrary to this, Meadowcraft compares social media to the image of Dorian Gray, which severs the link between action and consequence.[63] Social media's emphasis on emotional content, its asynchronous (and yet immediate) nature, and its nonymous nature can sever the bond between action and consequence. Vaidhyanathan says, "Facebook amplifies content

57. Meadowcroft, "The Distance Between Us," 78.
58. Meadowcroft, "The Distance Between Us," 78.
59. Cocking and Matthews, "Unreal Friends," 228.
60. Victorino, "Embracing Asynchronous Communication in the Workplace."
61. Meadowcroft, "The Distance Between Us," 78.
62. Meadowcroft, "The Distance Between Us," 78.
63. Meadowcroft, "The Distance Between Us," 74.

that hits strong emotional registers, whether joy or indignation."[64] Haidt and Rose Stockwell say the difference between the virtual and real worlds is that "[i]f you constantly express anger in your private conversations, your friends will likely find you tiresome, but when there's an audience, the pay-offs are different—outrage can boost your status."[65]

One result of this lack of connection between action and consequence is the development of what might be called a shame culture. Anton Barba-Kay says the "disparity between anonymity and publicity online is well captured by the senses of 'shame' in online contexts. The internet is—as a faceless medium, as a medium of seeing without being seen—essentially a shameless medium."[66] Barba-Kay continues, writing, "Identity becomes unmoored from some more or less particular audience in space and time, from the common settings that have always defined us, it becomes more dif-ficult to imagine ourselves as being answerable to and formed by some fixed group or community."[67] The result, he says, is poles of "shame and shame-lessness," forming the foundation of an "outrage industry" that "forcefully claims our public life."[68]

This culture of shame rests largely on trolls, who answer every claim with outlandish counters, which Nolen says mock "the opposing side for their hypocrisy while we nevertheless ignore or deny the hypocrisy of our own side."[69] This war of shame results in a denial of the humanness of those who oppose our views rather than mere disagreement. According to Nolen, "To say that we are Good, regardless of what we do, is to deny that we are what we do, and is thus to deny the nature of our own existence. To say that they are Evil, regardless of who they are, is to deny the humanity of those we oppose, and is thus to deny the nature of their existence."[70]

While several counterarguments have been raised to minimize, or even discount, the damage of virtual friendships, they fail when considered against the realities of identity formation and the flattening effects of neu-rodigital media. For instance, Søraker agrees virtual friendships are not full; they do not reduce happiness because they do not replace in-person friend-ships.[71] Research by R. I. M. Dunbar, however, shows social media does not

64. Vaidhyanathan, *Antisocial Media*, 8.

65. Haidt and Rose-Stockwell, "The Dark Psychology of Social Networks," 2.

66. Barba-Kay, "The Sound of My Own Voice," 6.

67. Barba-Kay, "The Sound of My Own Voice," 5.

68. Barba-Kay, "The Sound of My Own Voice," 5.

69. Gertz, *Nihilism and Technology*, loc. 3340.

70. Gertz, *Nihilism and Technology*, loc. 3340.

71. Søraker, "How Shall I Compare Thee?," 212.

increase the total number of relationships any person has, implying any deeply held relationships formed purely through social media must, in fact, replace an in-person relationship.[72]

Nicolas Munn agrees with the assessment that real friendship must include shared experiences, but then argues multiplayer games provide just such a shared experience through which real friendships can form.[73] Munn's argument, however, disregards the self-flattening and other-flattening effects of virtual worlds; individuals are not sharing a space together as themselves, but rather as some self-selected (and other-selected, via the capabilities and filtering modes of the environment the operator has created) version of themselves. Authentic self-disclosure, such as described by Cocking and Matthews, cannot take place in this kind of virtual environment.

The Commodification of Relationships

The gig economy is perhaps a perfect example of the commodification of relationships. Someone with a car, for instance, can contract (or work for) Uber or Lyft, which connect riders (customers) to drivers through an app on a smartphone. TaskRabbit is a similar system, only the services offered are personal shopping, doing general housework, running errands, and even what would have once been considered "handyman" work—light repairs, painting, and other tasks in support of maintaining a house. Another term, the *sharing economy*, is often used for services like Uber, but also for services like Airbnb, which allows homeowners to rent space in their homes for travelers to use instead of hotel rooms. While it seems these services merely allow people to sell what they already have, including their time and resources, Susie Armitage says, "Gig-economy apps promised to revolutionize the future of work. But they're also changing our concept of community."[74] Armitage notes that TaskRabbit began as a way to allow people to help one another with daily tasks—if someone who lives locally were already in a store, for instance, they could pick up a few items and drop them off for a little extra cash. TaskRabbit can be seen as a natural reaction to the fragmenting of suburban culture. In 2015, Joe Cortright reported among those he surveyed, "nearly a third report no interactions with neighbors and only about 20 percent say they spend time regularly with neighbors."[75] The reason for such services seems immaterial to Armitage's

72. Dunbar, "Do Online Social Media Cut Through the Constraints?"
73. Munn, "The Reality of Friendship Within Immersive Virtual Worlds," 5.
74. Armitage, "Gig-Economy Apps Are Changing Friendship."
75. Cortright, "Less in Common," 6.

concern about their impact on asking for favors, which "make us feel like part of a community."[76] She says that "relying on others is what holds us together—in families, romantic relationships, friendships, and as a society at large."[77]

The use of apps to pay for actions that once would have bound a community together is also the outcome of the differences in friction between the virtual and real (offline) worlds. Interacting with people in real life is rife with compromise over disagreements; the person on the other side of the table (or fence) is a *person* who demands respect and deference, who must be listened to, and whose opinions must be heard rather than simply shouted down or ignored. In the virtual world, a user can simply block another user with whom they consistently disagree, or even worse "unfriend" them. Barba-Kay writes,

> It is not that we are in danger of forgetting that we have parents, neighbors or classmates in the real world—these are on the whole still the most vivid and patent markers of who we are. But to the extent that online forms of sociability become more important to us, to the extent that our self-understanding is shaped by those forms, it is because we are bound to be always tempted to a place where we may avoid the inevitable frictions of offline relationships. Our online lives unfold, in contrast, within a safe space to which and from which we may always retreat to do our will.[78]

Sherry Turkle describes "companionship" with a robot as appearing to be "risk-free." Once a person has adapted to this kind of companionship, "life with people may seem overwhelming."[79]

The second aspect of commodification in relationships is quantification or rating every relationship in some way. In relation to the sharing or gig economy, quantification is the formal process of rating the service and the customer. Kari Paul argues that the shift from the consumer rating the business to a two-way rating system—where the business rates the customer as well—has created "equal power on both sides."[80] As noted earlier, Paul tells of a time when he rented an Airbnb in Brazil, and his host invited him to eat a traditional snack on his arrival. While the author was tired and just wanted to sleep, he felt obligated to eat with his host so he would be rated

76. Armitage, "Gig-Economy Apps Are Changing Friendship."
77. Armitage, "Gig-Economy Apps Are Changing Friendship."
78. Barba-Kay, "The Sound of My Own Voice," 5.
79. Turkle, *Alone Together*, 66.
80. Paul, "Rating Everything."

well, which would allow him to rent in other cities in the future.[81] In other cases, the sharing economy puts the consumer in direct contact with the provider, making the rating process complex; a bad rating might cost the provider their ability to continue working, for instance, or cause long-term adverse effects in their lives. Russell W. Belk writes, "People are playing a game and pretending it's a lovely social exchange—hosts sharing local tips with guests and guests sharing knowledge and skills from their native land—when it is, in fact, a business exchange."[82] As Gemma Newlands and coauthors say, "Rating mechanisms condition consumers toward perform-ing socially desirable behaviors during sharing transactions," and con-sumers often find such systems coercive.[83] This blurring of lines between personal and commercial relationships creates a mind-set of considering all relationships, no matter how personal, in terms of their instrumental value.

A third aspect of the quantification of relationships is the application of metrics by operators of social media and other services built on neurodigital media. Grosser points out that what "Facebook metrics want is increased user engagement with the site" because this drives profit through advertis-ing and selling information about users.[84] The result is what Grosser calls a "graphopticon," a situation where everyone is both watcher and watched, and everyone is rating everyone else through "likes," "reshares," and similar mechanisms. Gertz argues that this constant rating encourages "not shar-ing, not caring, but judging and discriminating."[85] He continues,

> It is not community that is the aim here but superiority, the superiority, first, of being a sharer, of being a crowdfunder, of choosing the moral path of putting people over profit, and second, the superiority of being able to accept and reject, the superiority of having been accepted and not rejected, the supe-riority of having a higher rating than others who have also been accepted. And because what these apps encourage runs counter to the ideology they promote, one feels not only superior but also guilty for having felt superior, for having felt the pleasure of pursuing power.[86]

Luke Fernandez and Susan Matt explain this kind of environment places the user in a position of treating others as an audience by which

81. Paul, "Rating Everything."
82. Cited in Paul, "Rating Everything."
83. Newlands, Lutz, and Fieseler, "The Conditioning Function," 1.
84. Grosser, "What Do Metrics Want?"
85. Gertz, *Nihilism and Technology*, loc. 2481.
86. Gertz, *Nihilism and Technology*, loc. 2481.

they can "have an identity" through the affirmation of others.[87] The authors continue, "Social media heightened self-consciousness and magnified social need."[88]

Environments designed to optimize quantification are depersonalized and damaging to relationship formation and maintenance. Barba-Kay points out that in these environments, "we may often deal with others whom we will never have to face up to, and that this intrinsic shamelessness affects how we are likely to respond to them."[89] He relates this to Godwin's Law, formulated early in the era of Internet chat rooms and bulletin-board systems, which states, "As an online discussion continues, the probability of a reference or comparison to Hitler or Nazis approaches one."[90]

Conclusion

Neurodigital media was designed to bring people together—to empower and create community by allowing people to share with one another without friction. The reality, however, seems far different. The filter bubble prevents those who want to speak from being heard and the spiral of silence prevents those outside the mainstream form speaking at all.

Virtual friendships are not equivalent to real ones. These virtual relationships can give the illusion of being in control—as Turkle says, "Our machine dream is to be never alone but always in control. This can't happen when one is face-to-face with a person. But it can be accomplished with a robot or . . . by slipping through the portals of a digital life."[91] The virtual world takes away the giving nature of relationships, replacing it with performance-driven quantification of relationships for "more."

Fernandez and Matt tell of a young man who walks across his college campus staring at his phone, although he is not using social media or talking to anyone. Instead, he uses the phone as a sort of shield against feeling left out. The phone, they say, "propped up the myth of individualism. It allowed one to appear independent, but simultaneously kept one from seeming too alone, for onlookers might assume that Skyler, standing by himself but staring at his phone, was connected to online friends, that he was popular on social media, if not on the quad."[92]

87. Fernandez and Matt, *Bored, Lonely, Angry, Stupid*, 68.

88. Fernandez and Matt, *Bored, Lonely, Angry, Stupid*, 68.

89. Barba-Kay, "The Sound of My Own Voice," 6.

90. Barba-Kay, "The Sound of My Own Voice," 6.

91. Turkle, *Alone Together*, 156.

92. Fernandez and Matt, *Bored, Lonely, Angry, Stupid*, 133.

10

Conclusion

THE PROMISE OF SOCIAL media—and all other systems built using neu-rodigital media—is, as William Powers says, to "erase the possibility of be-ing out of touch. To merge, to live with everyone, sharing every moment, every perception, thought, and action via our screens."[1] There is never any reason to miss a message from someone you know (or do not know). As Fernandez and Matt state, the meaning of being emotionally fulfilled in the digital age is "never being lonely, always being engaged and affirmed by others, being unconstrained in anger, and able to multitask and apprehend everything," to live without limits.[2]

In world of neurodigital media, there is also no reason to ever take the wrong turn, to miss a sale, or to not see the latest news about our friends. The consequences, however, are to never find and explore a new place by taking that wrong turn, to never have enough time to enjoy what we have because we are always seeking some new thing, and to never have time to spend with friends because we are always learning new things about our friends. These technologies have crept into everyday use without asking the essential question: what is gained, and what is lost.

For instance, Joseph Bottum describes an encounter with a computer-lab director epitomizing the attitude most users have towards these technol-ogies; addiction simply is not a problem because the technology is useful.[3] Several studies echo this sentiment; Amy Orben and Andrew Przybylski

1. Powers, *Hamlet's BlackBerry*, loc. 284.
2. Fernandez and Matt, *Bored, Lonely, Angry, Stupid*, 19.
3. Bottum, "Forty Years of the Computer Revolution," 1–2.

argue that the impact on adolescent well-being from digital technology use is "too small to warrant policy change."[4]

Why are we not asking the hard questions? Perhaps because our modern world is, as Amanda Mull observes it, miserable. She describes the products at a yearly technology exhibit as being made for people who are

> poorly rested, anxious about what they're eating, scared for the safety of their aging parents, and alienated from the natural responses of their body to things such as food and physical activity. They need to be stressed out and under pressure at work, not spending as much time as they'd like with family and friends, and unsure if they're doing the basics of modern human life— walking, sleeping, washing their face—correctly.[5]

Mull says, "For these devices to have any future at all, people have to be pretty miserable."[6] Modern life does seem to be miserable. Frank Ninivaggi reports there is an epidemic of loneliness.[7] Maybe the promise of a way out through neurodigital media is just too great to ignore.

Part of the Problem

Or maybe these ubiquitous systems built using neurodigital media are a part of the problem. Perhaps these systems cause us to be lonely and depressed, and by consuming our attention leave us no time to enjoy the people we are with and the things we already have.

Depression and Loneliness

Fernandez and Matt trace the concept of loneliness first to the Protestant Reformation and Calvinism, which had "fostered loneliness and implied it was inevitable,"[8] by rejecting the collective actions and the community of the Roman Catholic Church.[9] The Reformation, according to Max Weber, led to "unprecedented inner loneliness of the single individual," because for the person's "most important thing in life, his eternal salvation, he was forced to follow his path alone to meet a destiny which had been decreed

4. Orben and Przybylski, "The Association," 5–6.
5. Mull, "Gadgets for Life on a Miserable Planet."
6. Mull, "Gadgets for Life on a Miserable Planet," 2.
7. Ninivaggi, "Loneliness."
8. Fernandez and Matt, Bored, Lonely, Angry, Stupid, 102.
9. Fernandez and Matt, Bored, Lonely, Angry, Stupid, 85.

for him from eternity."[10] Contrary to Weber, the Protestant faith does offer
community, encouraging a personal relationship with a God who saves and
a horizontal relationship with communities of faith. But once naturalism
becomes the underlying worldview, as it is with current implementations of
neurodigital media, even the God who saves is gone, leaving the individual
standing alone before whatever powers of the world they might encounter.
Fernandez and Matt note a widespread reaction to this condition of being
alone was "New Thought," which taught "individuals could perfect themselves and improve their conditions through positive thinking," eliminating
loneliness by merely making themselves more likable.[11]

Marketers capitalized on this trend, portraying people who use technology to connect to others as successful and implying those who do not
were failures. Fernandez and Matt provide examples, such as the telephone
advertisement reading "every little while some friend or neighbor has a Bell
Telephone put in. If you have one, every new subscriber enlarges the scope
of your personal contact. If you have not, every new telephone makes you
the more isolated."[12] The authors say this has been extended into digital
systems, because "the idea that one can never have enough connections,
enough friends, enough opportunities to reach out and meet new people"
is implicit in today's tech marketing.[13] According to Fernandez and Matt,
this has led to a shift in the perception of the experience of being alone:
"Americans have become far less tolerant of boredom and dullness over the
last century, particularly so over the last decade. Whereas in the nineteenth
century, people were resigned to experiencing tedium and monotony in
their lives, contemporary Americans are primed for constant change, nov-
elty, and excitement."[14]

Several studies bear out these observations on social connections,
loneliness, and well-being in connection with neurodigital media. A study
by Wilson and coauthors shows how modern person is so attuned to being socially connected that they would rather shock themselves than be left
alone with their thoughts for an extended amount of time.[15] Taylor Wickel
performed a survey of social media users in the Millennial age group and
found they believe social media services are essential to their social life, and
that they base their view of the social status of the number of reactions to

10. Weber, *The Protestant Ethic and Spirit of Capitalism*, 104.

11. Fernandez and Matt, *Bored, Lonely, Angry, Stupid*, 102.

12. Fernandez and Matt, *Bored, Lonely, Angry, Stupid*, 104.

13. Fernandez and Matt, *Bored, Lonely, Angry, Stupid*, 128.

14. Fernandez and Matt, *Bored, Lonely, Angry, Stupid*, 140.

15. Wilson et al., "Just Think."

their posts on social media.[16] The author says, "As a way to acquire the maximum number of likes, Facebook users will manipulate and change their profile content. This is an indicator of narcissism in that respondents of the survey partake in an incessant need to pursue adoration from others, and to participate in egotistical thinking and behavior."[17] Ofir Turel, studying the social habits of younger users, shows there is a "decline of face-to-face social activities in youth that parallels the increase in the use of leisure technologies," including attending social functions, meeting friends, and going on dates.[18] Turel also documents a "general decline in well-being and selfworth perceptions that parallels the increase in leisure technology use."[19] Victoria Rideout and Michael Robb, in a study of social media's effect on social life, write,

> The proportion of teens who say their favorite way to communicate with their friends is "in person" has dropped from nearly half (49 percent) in 2012 (when it was their top choice) to less than a third (32 percent) today (when it's a close second to texting). And teens are more likely to say they're distracted from personal relationships by social media today than they were in 2012: Fifty-four percent of teens agree that using social media "often distracts me when I should be paying attention to the people I'm with," up from 44 percent in 2012; and 42 percent agree that the time they spend using social media "has taken away from time I could be spending with friends in person," up from 34 percent six years ago.[20]

Morten Tromholt designed a study where students agreed not to use Facebook for one week. He found increases in well-being among his subjects were directly related to the degree to which they relied on the service for their social connections.[21] Tromholt also found increases in the perception of well-being were related to what he called "Facebook envy" or the habit of the user comparing themselves to other users on the service.[22]

Tromholt's work echoes the work of Mai-Ly Steers and company, which found depressive symptoms were related to the act of social comparison

---

16. Wickel, "Narcissism and Social Networking Sites," 9.

17. Wickel, "Narcissism and Social Networking Sites," 9.

18. Turel, "Potential 'Dark Sides' of Leisure Technology Use in Youth," 25–26.

19. Turel, "Potential 'Dark Sides' of Leisure Technology Use in Youth," 26.

20. Rideout and Robb, "Social Media, Social Life."

21. Tromholt, "The Facebook Experiment," 664.

22. Tromholt, "The Facebook Experiment," 664.

with other users on social media services.[23] Brian Primack and his cohorts show increasing use of social media is correlated with increasing levels of perceived social isolation, saying those who visited social media services more than fifty-eight times per week "had about triple the odds of increased" perceived social isolation.[24] Melissa Hunt and her fellow researchers also performed a study where individuals were experimentally limited from using social media services for three weeks; the result was a "significant impact on well-being. Both loneliness and depressive symptoms declined in the experimental group."[25] They quote one participant as saying, "Not comparing my life to the lives of others had a much stronger impact than I expected, and I felt a lot more positive about myself during those weeks."[26] Elroy Boers also found a connection between time spent using digital devices and feelings of depression.[27]

Primack and coauthors speculated the correlation between social media use and perceived isolation could be related to the displacement of in-person experiences or the fear of missing out,[28] which Lovink describes as watching while

> others have rewarding experiences from which you are absent. That's the Fear of Missing Out, resulting in a constant desire for engagement with others and the world. This jealous feeling is the shadow side of the desire to be in the tribe, at the party, breast-to-breast. They dance and drink, while you're out there, on your own, in the cold. There is also another aspect: the online voyeurism, the detached form of peer-to-peer surveillance culture that carefully avoids direct interaction. Online we watch, and are watched. Overwhelmed by a false sense of familiarity with the Other, we're quickly bored and feel the urge to move on.[29]

The fear of missing out can manifest itself in a fear of not being able to use a physical device, such as a smartphone, through which the system built using neurodigital media is accessed. For instance, Stefan Tams and company demonstrated that users unable to access their smartphones during

23. Steers, Wickham, and Acitelli, "Seeing Everyone Else's Highlight Reels," 728.

24. Primack et al., "Social Media Use and Perceived Social Isolation," 3.

25. Hunt et al., "No More FOMO," 763.

26. Hunt et al., "No More FOMO," 751.

27. Boers et al., "Association of Screen Time and Depression," 853.

28. Primack et al., "Social Media Use and Perceived Social Isolation," 6.

29. Lovink, *Sad by Design*, loc. 835.

meetings perceived this lack of access as a social threat, calling the condition "Nomophobia."[30]

Beyond the fear of missing out, Turkle associates the loneliness of connection through virtual worlds to their flattening aspects, as discussed in chapter 3. She describes a young woman, Hannah, who is "able to imagine Ian as she wishes him to be. And he can imagine her as he wishes her to be. The idea that we can be exactly what the other desires is a powerful fantasy."[31] The relationship between this young woman and her partner is not with a person, but rather with their self-presentation through a social media service. While a relationship appears to be formed, it is not really (or fully) formed, leaving both the participants in a lonely kind of relationship, not quite a real relationship, but not quite alone. Turkle says, "Feeling secure as an object of desire (because the other is able to imagine you as the perfect embodiment of his or her desire) is one of the deep pleasures of Internet life."[32] Alan Jacobs argues that the virtual world engenders loneliness because it fails to manage "intimacy gradients: people using them always seem to feel either isolated or overwhelmed by crowds."[33] This isolation versus overwhelmed paradox is undoubtedly heightened by knowing that even in the most private of private chats, the operator is still observing user behavior, recording, and analyzing it for monetizable behavioral surplus.

Finally, many users fear the perceptions others will have of them based on their posts to social media. Wickel reports that nearly all the users in their study "determine another person's popularity based on how many 'likes' or comments that person's profile picture or status update receives."[34] In extreme cases, reactions to posts on social media services can result in job loss and even end a career. Kay says, "The idea that a whole career can fall victim to a single social-media message sent in a moment of anger or frustration—or even a bad joke—has produced an atmosphere of real terror."[35]

Lovink argues that the "evidence that sadness today is designed is overwhelming."[36] The design of these systems is to leave users in an ambiguous state, seeking ever greater amounts of attention. Lovink calls it "social hoovering," and says, "We're sucked back in, motivated by suggestive improvements in conditions that never materialize. Social media

---

30. Tams, Legoux, and Léger, "Smartphone Withdrawal," 2.

31. Turkle, *Alone Together*, 248.

32. Turkle, *Alone Together*, 248.

33. Jacobs, "Attending to Technology," 25.

34. Wickel, "Narcissism and Social Networking Sites," 8.

35. Kay, "The Tyranny of Twitter."

36. Lovink, *Sad by Design*, loc. 724.

architectures lock us in, legitimated by the network effect that everyone is on it—at least we assume they must be."[37] He continues, "Sadness expresses the growing gap between the self-image of a perceived social status and the actual precarious reality. The temporary dip, described here under the code name 'sadness,' can best be understood as a mirror phenomenon of the self-promotion machine."[38]

Consuming Attention

Josh Hawley argues, "Social media only works as a business model if it consumes users' time and attention day after day after day," creating what he calls an "attention arbitrage market."[39] Attention is deeply important to humans because, as James Williams says, "In order to do anything that matters, we must first be able to give attention to the things that matter."[40] Williams continues, discussing digital devices:

> [T]hese little wondrous machines, for all their potential, have not been entirely on our side. Rather than supporting our intentions, they have largely sought to grab and keep our attention. In their cutthroat competition against one another for the increasingly scarce prize of "persuading" us—of shaping our thoughts and actions in accordance with their predefined goals—they have been forced to resort to the cheapest, pettiest tricks in the book, appealing to the lowest parts of us, to the lesser selves that our higher natures perennially struggle to overcome.[41]

Rideout and Robb point out that nearly "three out of four teens (72 percent) believe that tech companies manipulate users to spend more time on their devices," with 54 percent of users reporting that the presence of electronic devices prevent them from paying attention to people present with them.[42] Fernandez and Matt point out that the operators' goals are not those of the user, saying, "Instead of your goals, success from their perspective is usually defined in the form of low-level 'engagement' goals, as they're often called. These include things like maximizing the amount of time you

37. Lovink, *Sad by Design*, loc. 755.
38. Lovink, *Sad by Design*, loc. 1045.
39. Hawley, "The Big Tech Threat."
40. Williams, *Stand Out of Our Light*, xi.
41. Williams, *Stand Out of Our Light*, xi.
42. Rideout and Robb, "Social Media, Social Life," 6.

spend with their product, keeping you clicking or tapping or scrolling as much as possible, or showing you as many pages or ads as they can."[43]

This attention economy, according to Mark Mason, has developed because the global Internet has made more information available "than any of us could possibly know what to do with."[44] The veritable glut of information transfers economic importance from what is being said to who is paying attention. Michael Goldhaber says attention "really is scarce, and the total amount per capita is strictly limited. To see why, consider yours, right now. It's going to these words. No matter how brilliant or savvy at multitasking you are, you can't be focusing on very much else. Ultimately then, the attention economy is a zero-sum game. What one person gets, someone else is denied."[45] He continues, "Attention can ground an economy because it is a fundamental human desire and is intrinsically, unavoidably scarce. It can be a rich and complex economy because attention comes in many forms: love, recognition, heeding, obedience, thoughtfulness, caring, praising, watching over, attending to one's desires, aiding, advising, critical appraisal, assistance in developing new skills, et cetera."[46]

Fernandez and Matt say that, in this attention economy, "the chief means for making profits is through distraction. Because these distractions have become integral to the business process, they are at once vastly more ubiquitous and much more entrenched."[47] Because the mental space of each person is bound both by time and space, attention can only be drawn to an individual piece of information for a particular period of time. According to Matthew Crawford, however, attention is so precious "we've auctioned off more and more of our public space to private commercial interests, with their constant demands on us to look at the products on display or simply absorb some bit of corporate messaging. . . . In the process, we've sacrificed silence—the condition of not being addressed. And just as clean air makes it possible to breathe, silence makes it possible to think."[48]

Constant distraction results in information overload. While Varvara Chumakova says, "the human brain usually works in a state of information overload," the brain naturally filters out information deemed unimportant in some way.[49] Chumakova notes, however, that "information overload

43. Williams, *Stand Out of Our Light*, 8.

44. Manson, "In the Future, Our Attention Will Be Sold."

45. Goldhaber, "Attention Shoppers!"

46. Goldhaber, "Attention Shoppers!"

47. Fernandez and Matt, *Bored, Lonely, Angry, Stupid*, 240.

48. Crawford, "The Cost of Paying Attention."

49. Chumakova, "Surviving Information Overload," 1.

transforms people's attitudes towards themselves and others, for it facili-
tates borders between a person and his or her locality."[50] According to Carr,
"The influx of competing messages that we receive whenever we go online
not only overloads our working memory; it makes it much harder for our
frontal lobes to concentrate our attention on any one thing. The process of
memory consolidation can't even get started."[51] Carr says this results in a
feedback loop where using the Internet extensively trains the brain to be
distracted, causing shallower thinking, and thus making the individual ever
more dependent on the "easily searchable artificial memory" of the Internet.[52] Williams describes the effect as "los[ing] control over one's attentional
processes. In other words, the problems in Tetris arise not when you stack a
brick in the wrong place (though this can contribute to problems down the
line), but rather when you lose control of the ability to direct, rotate, and
stack the bricks altogether."[53]

Mark Manson notes the importance of self-control in an attention
economy, saying, "In the new economy, the most valuable asset you can
accumulate may not be money, may not be wealth, may not even be knowledge, but rather, the ability to control your own attention, and to focus."[54]
Rousseau writes, "We might, over and above all this, add, to what man ac-
quires in the civil state, moral liberty, which alone makes him truly master
of himself; for the mere impulse of appetite is slavery, while obedience to
a law which we prescribe to ourselves is liberty."[55] Following up on Rousseau, Williams says, "The liberation of human attention may be the defining
moral and political struggle of our time. Its success is prerequisite for the
success of virtually all other struggles."[56]

Reviewing the Argument

Neurodigital media is rapidly integrating into individual lives and the
broader culture in the form of social media networks like Facebook and
LinkedIn, social recommender systems used to guide individuals through
their purchases on sites such as Amazon, and into other widely used systems. Understanding the impact these systems have on individuals and

50. Chumakova, "Surviving Information Overload," 1.

51. Carr, The Shallows, loc. 3267.

'52. Carr, The Shallows, loc. 3297.

53. Williams, Stand Out of Our Light, 15.

54. Manson, "In the Future, Our Attention Will Be Sold."

55. Rousseau, The Social Contract & Discourses, 19.

56. Williams, Stand Out of Our Light, xii.

culture requires exposing the vision of the person they embody and how that vision is expressed in their design and operation. The sections below trace the broad contours of the argument presented in this book to emphasize the logical steps from systems built on neurodigital media to the resulting harm to persons within a Christian anthropology.

Neurodigital Media and Naturalism

Technology, however, does not stand alone; it must embody some worldview, some vision of reality. Neurodigital media primarily arises out of an area and culture commonly known as Silicon Valley, in reference to a geographic area around between the Diablo and Santa Cruz mountains extending south from San Francisco to Cupertino, California.[57] In this area, according to Fred Turner, the counterculture of the 1960s mixed with the engineering culture of the military-industrial complex.[58] The resulting culture is deeply progressive, holding the future can be shaped by humans to reach an ideal state. J. B. Bury argues that this culture is effectively a newly created religion.[59] Douglas Rushkoff says this religion is a "techno-utopian and deeply anti-human sensibility, born out of a little-known confluence of American and Soviet New Age philosophers, scientists, and spiritualists who met up in the 1980s hoping to prevent nuclear war—but who ended up hatching a worldview that's arguably as dangerous to the human future as any atom bomb."[60]

Richard Barbrook and Andy Cameron, in an early and influential article, call this culture the *Californian Ideology*, and say it is "naturalising and giving a technological proof to a libertarian political philosophy," through a "mix of cybernetics, free market economics, and counter-culture libertarianism."[61] The progressive urge to reshape humans and society is supported by naturalistic anthropology, which holds humans are, according to Joel Green, "nothing but the product of organic chemistry."[62] Combining the urge to reshape individuals towards some imagined vision of perfection with naturalistic anthropology, according to C. S. Lewis, means that "what

57. Dennis, "Silicon Valley."
58. Turner, *Counterculture to Cyberculture*, loc. 181.
59. Bury, *The Idea of Progress*, 286.
60. Rushkoff, "The Anti-Human Religion of Silicon Valley."
61. Barbrook and Cameron, "The Californian Ideology."
62. Green, *Body, Soul, and Human Life*, 30.

we call Man's power over Nature turns out to be a power exercised by some men over other men with Nature as its instrument."[63]

Naturalistic and Christian anthropology contrast in three specific areas: moral freedom, dignity, and the purpose of relationships. In naturalism, all three of these are the result of evolutionary and emergent processes, so each has only an instrumental purpose. According to Richard Dawkins, belief in naturalism commits one to holding that free will is a helpful evolutionary adaptation that simply does not exist.[64] C. S. Lewis agrees, stating, "No thoroughgoing Naturalist believes in free will."[65] In contrast, Christian anthropology holds that moral freedom is an essential property of humans, according to Christian thinkers such as Michael Novak[66] and Dale Patrick.[67] R. T. Kendall lists nine different passages in the Scriptures indicating individuals have moral freedom.[68] Some of these are familiar, such as Joshua 24:15, of which Kendall says, "Had Israel been incapable of making the right choice, Joshua is mocking them in requiring such a choice."[69] He also argues that Elijah, in 1 Kings 18:21, assumes individuals can choose to worship God or Baal.[70] Kendall further notes that the call to repentance assumes individuals are able to make morally significant choices, using Acts 17:30 and 2:38 as examples, and repentance and conversion to Christian faith is "the result of being persuaded."[71] For this latter example, Kendall cites 2 Corinthians 5:11 and Acts 26:28. He also holds that the words of Christ concerning Israel's rejection of the Messiah in Luke 19:41 and Matthew 23:37 indicate the ability of individuals to freely make morally significant choices.[72]

Dignity is the second area where Christian and naturalistic anthropology deeply differ. Naturalistic anthropology often grounds dignity in the right to self-determination. Ron Highfield says, "Treat[ing] humans according to their dignity is to treat them as if they owned themselves and had a right to determine their own actions."[73] Grounding dignity in moral

---

63. Lewis, *The Abolition of Man*, 55.

64. *FREE WILL—Lawrence Krauss and Richard Dawkins.*

65. Lewis, *Miracles*, 7.

66. Novak, "The Judeo-Christian Foundation of Human Dignity," 112.

67. Patrick, "Studying Biblical Law as a Humanities," 34.

68. Kendall, *Understanding Theology*, 2:278.

69. Kendall, *Understanding Theology*, 2:278.

70. Kendall, *Understanding Theology*, 2:278.

71. Kendall, *Understanding Theology*, 2:278.

72. Kendall, *Understanding Theology*, 2:278.

73. Highfield, *God, Freedom and Human Dignity*, loc. 1009.

freedom is problematic because, according to naturalistic anthropology, free will is an evolutionary adaptation rather than a real property of individuals. B. F. Skinner declares, "What is being abolished is autonomous man—the inner man, the homunculus, the possessing demon, the man defended by the literatures of freedom and dignity. His abolition has long been overdue."[74] Christian anthropology, on the other hand, grounds human dignity in the *imago Dei*, the image of God in every person, according to Stephen P. Greggo, Lucas Tillett,[75] and Carl F. H. Henry.[76]

The Scriptures provide wide-ranging support for the dignity of individuals in Christian anthropology; as H. C. Leupold says, "The true dignity of man is taught nowhere as effectively as in the Scriptures."[77] Charles M. Horne argues for the dignity of humans from Psalm 8:4–5, which says, "What is man that you are mindful of him, and the son of man that you care for him? Yet you have made him a little lower than the heavenly beings and crowned him with glory and honor."[78] Horne argues individual humans have dignity because "man alone was created in God's image."[79] Robert Gonzales, Jr., lists five places in the Scriptures where man is said to have the *imago Dei*: Genesis 1:26–27; 5:1–2; 9:6; 1 Corinthians 11:7; and James 3:9.[80] Gonzales relates the *imago Dei* directly to the dignity of the individual, holding that "Man is not a product that randomly (by chance) evolved from primates, as evolution teaches. If that were true, then man would have no dignity or meaning in life."[81]

Finally, naturalistic anthropology holds that relationships are important, but they are merely instrumental. Dennis L. Krebs states the human sense of morality arises as a counter to the unfairness of hierarchical communities, which in turn formed to control aggressive behavior.[82] Eric Alden Smith rules out teleological or supernatural origins when discussing the origins of human social behavior, saying explanations of relationship formation must be "functional accounts and ask how a trait of interest contributes to adaptive success."[83] In contrast, Christian anthropology holds that

---

74. Skinner, *Beyond Freedom and Dignity*, 200.

75. Greggo and Tillett, "Beyond Bioethics 101," 356.

76. Henry, *God, Revelation, and Authority*, 6:454.

77. Leupold, *Exposition of the Psalms*, 100.

78. Horne, "Christian Humanism," 186.

79. Horne, "Christian Humanism," 186.

80. Gonzales, Jr., "Man," 85.

81. Gonzales, Jr., "Man," 85.

82. Krebs, "Morality," 149.

83. Smith, "Agency and Adaptation," 107.

humans are designed to be in relationship—that individuals cannot truly flourish without being in a community. Anthony Hoekema says, "The image must be seen in man's threefold relationship: toward God, toward others, and toward nature. When originally created, humans imaged God sinlessly in all three relationships."[84]

The Christian view of relationships is directly supported within the Scriptures beginning with the creation narrative in the second chapter of Genesis. In Genesis 2:19–20, God brings "every beast of the field and every bird of the heavens" to Adam to name. In the end, however, "there was not found a helper fit for him." To remedy this situation, God took a portion from Adam's side to create the woman. On seeing the woman for the first time, Adam declares in Genesis 2:23, "This at last is bone of my bones and flesh of my flesh; she shall be called Woman, because she was taken out of Man." Genesis 2:24 continues, "Therefore shall a man leave his father and mother and hold fast to his wife, and they shall become one flesh." The importance of relationships is also emphasized by Jesus. After being asked what the most important commandment is, he answers in Matthew 22:37–9, "You shall love the Lord your God with all your heart and with all your soul and with all your mind. This is the great and first commandment. And a second is like it: You shall love your neighbor as yourself." Both of the two commandments on which Jesus says the entire law and prophets depend (Matthew 22:40) involve relationships. The first commandment speaks of the relationship between the individual and God; the second speaks of the relationship between individuals. Embodying the naturalistic, instrumental, view of moral freedom, dignity, and community into systems built using neurodigital media harms individual users from the perspective of Christian anthropology.

The Power of Neurodigital Media

Systems built using neurodigital media draw users into their ecosystem by creating a space where individuals can build an audience and perform—because the resonance of neurodigital media is performance. Once users are inside the ecosystem, they can be shaped by the immersive user experience. Richard Seymour calls these systems a form of the Skinner Box, saying,

> No one forces us to be there, and no one tells us what to post, 'like' or click. And yet our interactions with the machine are conditioned. Critics of social media like Jaron Lanier argue that

84. Hoekema, *Created in God's Image*, 95.

the user experience is designed much like the famous 'Skinner Box' or 'operant conditioning chamber' invented by the pioneering behaviourist B. F. Skinner. In this chamber, the behaviour of laboratory rats was conditioned by stimuli—lights, noises and food. Each of these stimuli constituted a 'reinforcement', either positive or negative, which would reward some forms of behaviour and discourage others. In the Skinner Box, test subjects are taught how to behave through conditioning. And if this model has found its way into the mobile apps, gaming and social media industries, it might reflect the way that behaviourist ideas have achieved a surprising renaissance among businessmen and policymakers in recent decades.[85]

The "like button" is the crux of this experience. The desire to build an audience, to have someone pay attention and belong to a virtual community, drives some users to invest large portions of their lives in these online services.

The desire to perform, combined with the user experience, draws users into engaging with the service more frequently and sharing ever-larger amounts of information. Ultimately, the operator attempts to reach a state of frictionless sharing, where the user reveals even the most trivial and intimate parts of their life without thinking. Anand Jagannathan describes frictionless sharing as minimizing the amount of work users must do to share, such as keeping them logged onto a service through a specialized app.[86] Services built using neurodigital media often go beyond frictionless sharing by surveilling users directly. For instance, services might track user location, listen to a user's surroundings through a device's microphone, and track the user's searches.

The information gained through this pervasive surveillance is analyzed through machine learning and artificial intelligence systems to determine what each user likes, is afraid of, and influenced by—a detailed profile of each user. Shoshana Zuboff says this process "unilaterally claims human experience as free raw material for translation into behavioral data," and "declared as a proprietary behavioral surplus."[87] These profiles can then be used to influence, or as Richard Thaler and Cass Sunstein say, give a nudge, which "alters people's behavior in a predictable way."[88] Ubiquitous surveillance, an immersive user experience iteratively designed using techniques

85. Seymour, *The Twittering Machine*, 23.
86. Jagannathan, "Frictionless Sharing."
87. Zuboff, *The Age of Surveillance Capitalism*, 7.
88. Thaler and Sunstein, *Nudge*, 10.

borrowed from operant conditioning, and the ability to take advantage of natural human desires for recognition combine to create systems with unparalleled power over user beliefs and actions.

The Impact of Neurodigital Media

Recognizing the power neurodigital media holds over its users, operators could put rules in place to contain its use within ethical bounds. The twin motives of profit and progressive impulses, however, have allowed operators to ignore ethical concerns applied elsewhere. As Zuboff says,

> Surveillance capitalists quickly realized that they could do anything they wanted, and they did. They dressed in the fashions of advocacy and emancipation, appealing to and exploiting contemporary anxieties, while the real action was hidden offstage. Theirs was an invisibility cloak woven in equal measure to the rhetoric of the empowering web, the ability to move swiftly, the confidence of vast revenue streams, and the wild, undefended nature of the territory they would conquer and claim. They were protected by the inherent illegibility of the automated processes that they rule, the ignorance that these processes breed, and the sense of inevitability that they foster.[89]

The naturalistic anthropology that neurodigital media embodies provides no ethical bounds to control the twin quests for profit and shaping societies towards progressive ends. On the contrary, naturalism strongly supports these motivations; the operator of a service based on neurodigital media can claim to help the world become a better place while at the same time maintaining extremely high profit levels.

Neurodigital media truncates human freedom by reducing uncertainty through nudging and intentionally habituating users to engaging through (or with) a service, sometimes to the point of addiction. Arielle Pardes documents a series of nudges built into social media interfaces classified as dark patterns, such as the options services like Facebook and Twitter present to users about "ad personalization."[90] Pardes gives an example: "A recent Twitter pop-up told users 'You're in control,' before inviting them to 'turn on personalized ads' to 'improve which ones you see' on the platform. Don't want targeted ads while doomscrolling? Fine. You can 'keep less

---

89. Zuboff, *The Age of Surveillance Capitalism*, 9.
90. Pardes, "How Facebook and Other Sites Manipulate Your Privacy Choices."

relevant ads.' Language like that makes Twitter sound like a sore loser."[91] There is no option to turn advertisements off, just the confusing option between making advertisements "more" or "less relevant"—with little explanation of what the difference between these options might be (other than making it clear "more relevant" is better for the user in some way).

Modifying user behavior is often justified by showing how it can improve individual lives or whole societies. For instance, Thaler and Sunstein call the nudge libertarian paternalism, writing,

> The paternalistic aspect lies in the claim that it is legitimate for choice architects to try to influence people's behavior in order to make their lives longer, healthier, and better. In other words, we argue for self-conscious efforts, by institutions in the private sector and also by government, to steer people's choices in directions that will improve their lives. In our understanding, a policy is "paternalistic" if it tries to influence choices in a way that will make choosers better off, as judged by themselves.[92]

This argument assumes an instrumental view of the person that fits well in naturalistic anthropology but is contrary to the Christian view of the person. Overriding the moral decision-making of individuals harms users by leaving them less able to make moral decisions. As C. S. Lewis says, "Each new power won by man is a power over man as well. Each advance leaves him weaker as well as stronger."[93]

Systems built using neurodigital media subvert human dignity through a process of flattening, or the abstraction of the user to a reduced set of attributes for classification and analysis. Users flatten themselves in constructing an online identity as a form of performance designed to build what Geert Lovink calls social capital.[94] Users also flatten one another through the quantification of relationships. Benjamin Grosser, for instance, argues that users quantify their relationships with others by seeking out approval through the number of likes, connections, and other metrics provided by the system.[95] Once again, the subversion of dignity by systems built on neurodigital media fits well within the naturalistic anthropology they embody, which treats each person as having primarily instrumental value. In contrast, Christian anthropology treats each person as having value because

91. Pardes, "How Facebook and Other Sites Manipulate Your Privacy Choices."
92. Thaler and Sunstein, *Nudge*, 8.
93. Lewis, *The Abolition of Man*, 55.
94. Lovink, *Sad by Design*, 19.
95. Grosser, "What Do Metrics Want?"

they bear the *imago Dei*. As Richard Lints says, "The dignity of the imago Dei was its ability to reflect and relate to the true and living God."[96]

Systems built using neurodigital media distort the social order and harm the formation of the kinds of community Christian anthropology demands for individual humans. These systems harm community formation by creating filter bubbles and chilling speech, ultimately reducing each community to a single person. Systems built using neurodigital media present a reduced form of friendship in which, according to Micah Meadowcroft, users interact with datums rather than one another.[97] These systems also intentionally consume the attention of their users to the exclusion of friends and family who are physically present.

The threefold impact on the person—truncating freedom, subverting dignity, and hampering community formation—leads to widespread depression and other social ills. According to several studies, such as one performed by Morten Tromholt, the use of these services also bring about a general lack of well-being up to and including depression.[98] From the perspective of Christian anthropology, the adverse effects are not surprising, as the naturalistic anthropology embodied in systems built using neurodigital media subvert three crucial attributes God designed into humans.

Concluding Thoughts

On reaching this point, readers might think there is no hope—these companies are too powerful, and the technology is too "good" at capturing attention, to do anything about this problem. As one technologist has said, "prepare to bow before your new overlords."[99]

While this is an understandable reaction, humankind has been on the cusp of new ages with technologies that appeared to threaten the very existence of society itself. Looking back in history, how have past generations handled these transitions? What forms of resistance, if that is the correct word, can the average person on the street use against these technologies?

Do not despair—there is hope, and there are possible paths forward. This book, however, has already grown too long, so thoughts around how Christians can effectively counter these technologies, remaining a witness to the world and providing the "salt" of a better path forward will need to wait until the next book.

96. Lints, *Identity and Idolatry*, 76.

97. Meadowcroft, "The Distance Between Us," 78.

98. Tromholt, "The Facebook Experiment," 664.

99. White, Ammon, and Huston, "The Hedge 6."

Bibliography

Aberdeen, J. A. "The Edison Movie Monopoly." Hollywood Renegades Archive, 2005. http://www.cobbles.com/simpp_archive/edison_trust.htm.

"About LinkedIn." https://about.linkedin.com/.

"About TikTok." https://www.tiktok.com/about?lang=en.

Achara, Jagdish Prasad, Javier Parra-Arnau, and Claude Castelluccia. "Fine-Grained Control over Tracking to Support the Ad-Based Web Economy." *ACM Transactions on Internet Technology* 18, no. 4 (September 2018) 51:1–51:25. https://doi.org/10.1145/3158372.

Acquisti, Alessandro, Curtis R. Taylor, and Liad Wagman. "The Economics of Privacy." *Journal of Economic Literature* 52, no. 2 (August 2016) 442–92. http://doi.org/10.1257/jel.54.2.442.

Adelhardt, Zinaida, Stefan Markus, and Thomas Eberle. "Teenagers' Reaction on the Long-Lasting Separation from Smartphones, Anxiety and Fear of Missing Out." In *Proceedings of the 9th International Conference on Social Media and Society*, 212–16. SMSociety '18. Copenhagen, Denmark: Association for Computing Machinery, 2018. https://doi.org/10.1145/3217804.3217914.

AdExchanger. "If A Consumer Asked You, 'Why Is Tracking Good?', What Would You Say?" *AdExchanger* (blog), October 2011. https://www.adexchanger.com/online-advertising/why-is-tracking-good/.

Alsaedi, Nasser, Pete Burnap, and Omer Rana. "Can We Predict a Riot? Disruptive Event Detection Using Twitter." *ACM Transactions on Internet Technology* 17.2 (2017) 18:1–18:26. https://doi.org/10.1007/s10676-016-9388-y.

Ali, Muhammad, Piotr Sapiezynski, Aleksandra Korolova, Alan Mislove, and Aaron Rieke. "Ad Delivery Algorithms: The Hidden Arbiters of Political Messaging." *ArXiv*, December 2019, 1–16. https://arxiv.org/abs/1912.04255.

Alter, Adam. *Irresistible: The Rise of Addictive Technology and the Business of Keeping Us Hooked*. Harmondsworth: Penguin, 2018.

"Amazon.com Link." https://www.amazon.com/Hidden-Persuaders-Vance-Packard-ebook/dp/B006NV977W/ref=tmm_kin_swatch_0?_encoding=UTF8&qid=1581841039&sr=8-1.

Anderson, Chris. "The End of Theory: The Data Deluge Makes the Scientific Method Obsolete." *Wired*, June 2008. https://www.wired.com/2008/06/pb-theory/.

Andrew-Gee, Eric. "Your Smartphone Is Making You Stupid, Antisocial and Unhealthy. So Why Can't You Put It Down?" *The Globe and Mail*, April 2018. https://www.

theglobeandmail.com/technology/your-smartphone-is-making-you-stupid/
article37511900/.

Android Developers. "Manifest.Permission." https://developer.android.com/reference/
android/Manifest.permission.

Anticoli, Linda, and Marco Basaldella. "Shut up and Run: The Never-Ending Quest
for Social Fitness." In *Companion Proceedings of the The Web Conference*,
1553–56. WWW '18. Association of Computing Machinery, 2018. https://doi.
org/10.1145/3184558.3191609.

Armitage, Susie. "Gig-Economy Apps Are Changing Friendship." *Slate Magazine*
(blog), January 2020. https://slate.com/technology/2020/01/gig-economy-apps-
are-changing-friendship.html.

Arora, Shabana. "Recommendation Engines: How Amazon and Netflix Are Winning
the Personalization Battle." *MarTech Advisor* (blog), June 2016. https://www.
martechadvisor.com/articles/customer-experience-2/recommendation-engines-
how-amazon-and-netflix-are-winning-the-personalization-battle/.

Arp, Daniel, Erwin Quiring, Christian Wressnegger, and Konrad Rieck. "Privacy
Threats through Ultrasonic Side Channels on Mobile Devices." In *2017 IEEE
European Symposium on Security and Privacy*, 35–47. https://doi.org/10.1109/
EuroSP.2017.33.

Aten, Jason. "Google Is Making It Harder to Tell the Difference Between Ads and
Search Results." *Inc.Com* (blog), January 2020. https://www.inc.com/jason-aten/
google-is-making-it-harder-to-tell-difference-between-ads-search-results.html.

Baime, A. J. *The Arsenal of Democracy: FDR, Detroit, and an Epic Quest to Arm an
America at War*. New York: Mariner, 2014.

Balakrishnan, Mahesh, Iqbal Mohomed, and Venugopalan Ramasubramanian.
"Where's That Phone? Geolocating IP Addresses on 3G Networks." In *Proceedings
of the 9th ACM SIGCOMM Conference on Internet Measurement*, 294–300.
IMC '09. Chicago: Association for Computing Machinery, 2009. https://doi.
org/10.1145/1644893.1644928.

Banton, Caroline. "Understanding Just-in-Time (JIT) Inventory Systems." Investopedia,
February 2020. https://www.investopedia.com/terms/j/jit.asp.

Barba-Kay, Anton. "The Sound of My Own Voice." *The Point Magazine*, February 2019.
https://thepointmag.com/2019/politics/the-sound-of-my-own-voice.

Barbrook, Richard, and Andy Cameron. "The Californian Ideology." *Mute*, September
1995. http://www.metamute.org/editorial/articles/californian-ideology.

Barlow, John Perry. "A Declaration of the Independence of Cyberspace." Electronic
Frontier   Foundation,   January   2016.   https://www.eff.org/cyberspace-
independence.

Barresi, John, and Raymond Martin. *Naturalization of the Soul: Self and Personal
Identity in the Eighteenth Century*. London: Routledge, 2000.

Barrett, William. *Death of the Soul: From Descartes to the Computer*. Oxford: Oxford
University Press, 1987.

Baumeister, R. F., and M. R. Leary. "The Need to Belong: Desire for Interpersonal
Attachments as a Fundamental Human Motivation." *Psychological Bulletin* 117,
no. 3 (May 1995) 497–529. https://doi.org/10.1037/0033-2909.117.3.497.

Baumer, Eric P. S., Shion Guha, Emily Quan, David Mimno, and Geri K. Gay. "Missing
Photos, Suffering Withdrawal, or Finding Freedom? How Experiences of Social

Media Non-Use Influence the Likelihood of Reversion." *Social Media + Society* 1, no. 2 (July 2015) 1–14. https://doi.org/10.1177/2056305115614851.

Beard, Charles. *The American Leviathan: The Republic in the Machine Age*. New York: Macmillan, 1930.

Belvedere, Benny. "Facebook Makes More If You're Addicted to the Internet and Fake News." *The Federalist* (blog), October 2017. http://thefederalist.com/2017/10/10/facebook-makes-money-youre-addicted-internet-fake-news/.

Benevenuto, Fabrício, Matheus Araújo, and Filipe Ribeiro. "Sentiment Analysis Methods for Social Media." *Proceedings of the 21st Brazilian Symposium on Multimedia and the Web*. WebMedia '15. Manaus, Brazil: Association for Computing Machinery, 2015. https://doi.org/10.1145/2820426.2820642.

Bentham, Jeremy. *The Works of Jeremy Bentham*. Edited by John Bowring. Vol. 4, 1843. https://oll.libertyfund.org/titles/bentham-the-works-of-jeremy-bentham-vol-4#lfo872–04_head_004.

Berreby, David. "Is It a Nudge or a Shove? Even Experts Can Blur the Lines." *Psychology Today* (blog), November 2016. https://www.psychologytoday.com/blog/the-outsourced-mind/201611/is-it-nudge-or-shove-even-experts-can-blur-the-lines.

Bhattacharyya, Suman. "Pressured by Amazon, Retailers Are Experimenting with Dynamic Pricing." *Digiday* (blog), February 2019. https://digiday.com/retail/amazon-retailers-experimenting-dynamic-pricing/.

Biddle, Sam, Paulo Victor Ribeiro, and Tatiana Dias. "Invisible Censorship: TikTok Told Moderators to Suppress Posts by 'Ugly' People and the Poor to Attract New Users." *The Intercept* (blog), March 2020. https://theintercept.com/2020/03/16/tiktok-app-moderators-users-discrimination/.

Bilal, Mohammed. "How to Design a Habit-Forming Shopping Experience." *FreeCodeCamp.Org* (blog), May 2017. https://medium.freecodecamp.org/how-to-design-a-habit-forming-shopping-experience-af7748402e90.

Bimber, Bruce. "Three Faces of Technological Determinism." In *Does Technology Drive History? The Dilemma of Technological Determinism*, edited by Merritt Roe Smith and Leo Marx, 79–100. Cambridge, MA: The MIT Press, 1994.

Boa, Kenneth D. "What Is Behind Morality?" In *Vital Contemporary Issues: Examining Current Questions and Controversies*, edited by Roy B. Zuck, 12–23. Grand Rapids: Kregel, 1994.

Boers, Elroy, Mohammad H. Afzali, Nicola Newton, and Patricia Conrod. "Association of Screen Time and Depression in Adolescence." *JAMA Pediatrics* 173, no. 9 (September 2019) 853–59. https://doi.org/10.1001/jamapediatrics.2019.1759.

Bogost, Ian. "Stop Trusting Viral Videos." *The Atlantic* (blog), January 2019. https://www.theatlantic.com/technology/archive/2019/01/viral-clash-students-and-native-americans-explained/580906/.

Bohay, Mark, Daniel P. Blakely, Andrea K. Tamplin, and Gabriel A. Radvansky. "Note Taking, Review, Memory, and Comprehension." *The American Journal of Psychology* 124, no. 1 (2011) 63–73. https://doi.org/10.5406/amerjpsyc.124.1.0063.

Boice, James Montgomery. *Romans: The New Humanity*. Vol. 4. Grand Rapids: Baker, 1991.

Bond, Robert M., Christopher J. Fariss, Jason J. Jones, Adam D. I. Kramer, Cameron Marlow, Jaime E. Settle, and James H. Fowler. "A 61-Million-Person Experiment in Social Influence and Political Mobilization." *Nature* 489, no. 7415 (September 2012) 295–98. https://doi.org/10.1038/nature11421.

Borysenko, Karlyn. "The Ultimate Crowdsourced Definition of a Karen in the Era of COVID-19." *Medium* (blog), May 2020. https://medium.com/@karlyn/the-ultimate-crowdsourced-definition-of-a-karen-in-the-era-of-covid-19-f1e53057879.

Bossetta, Michael. "Spear Phishing and Cyberattacks on Democracy." *Journal of International Affairs* 71, no. 1.5 (2018) 97–106. https://www.jstor.org/stable/26508123.

Botella, Elena. "Facebook Earns $132.80 From Your Data per Year." *Slate Magazine* (blog), November 2019. https://slate.com/technology/2019/11/facebook-six4three-pikinis-lawsuit-emails-data.html.

Bottum, Joseph. "Forty Years of the Computer Revolution." *The American Spectator* (blog), December 2019. https://spectator.org/forty-years-of-the-computer-revolution/.

Bouk, Dan. *How Our Days Became Numbered: Risk and the Rise of the Statistical Individual.* Kindle ed. Chicago: University of Chicago Press, 2015.

Bowles, Nellie. "Human Contact Is Now a Luxury Good." *New York Times*, March 2019. https://www.nytimes.com/2019/03/23/sunday-review/human-contact-luxury-screens.html.

Bradbury, Ray. *Fahrenheit 451.* New York: Simon & Schuster, 2012.

Brady, William J., Ana P. Gantman, and Jay Joseph Van Bavel. "Attentional Capture Helps Explain Why Moral and Emotional Content Go Viral." *Journal of Experimental Psychology: General* 148, no. 9 (September 2019) 1–63. https://doi.org/10.31234/osf.io/zgd29.

Brazier, P. H. *C. S. Lewis—On the Christ of a Religious Economy: I. Creation and Sub-Creation.* Vol. 3.1. C. S. Lewis: Revelation and the Christ. Eugene, OR: Pickwick, 2013.

Brodwin, Erin. "There's No Solid Evidence That People Get Addicted to Social Media—and Using It Could Actually Be Beneficial." *Business Insider* (blog), March 2018. https://www.businessinsider.com/social-media-iphone-facebook-instagram-addiction-2018-3.

Bruni, Ezequiel. "8 Ways to Emotionally Reward Your Users." *Webdesigner Depot* (blog), April 2018. https://www.webdesignerdepot.com/2018/04/8-ways-to-emotionally-reward-your-users/.

Bury, J. B. *The Idea of Progress: An Inquiry into Its Origin and Growth.* New York: Dover, 2014.

Byrne, Ciara. "Why Google Defined a New Discipline to Help Humans Make Decisions." *Fast Company*, July 2018. https://www.fastcompany.com/90203073/why-google-defined-a-new-discipline-to-help-humans-make-decisions.

Calo, Ryan. "Digital Market Manipulation." *George Washington Law Review* 82 (January 2014) 995–1051. https://digitalcommons.law.uw.edu/faculty-articles/25.

Camacho, Francisco Gómez, and Luis de Molina. "Treatise on Money." Translated by Jeannine Emery. *Journal of Markets & Morality* 8, no. 1 & 2 (2005) 161–325.

Carr, Nicholas. *The Shallows: What the Internet Is Doing to Our Brains.* Kindle ed. New York: Norton, 2011.

Carroll, Joseph. "Evolutionary Social Theory: The Current State of Knowledge." *Style* 49, no. 4 (2015) 512–41. https://doi.org/10.5325/style.49.4.0512.

Case, Anne, and Angus Deaton. *Deaths of Despair and the Future of Capitalism.* Princeton, NJ: Princeton University Press, 2020.

Chumakova, Varvara. "Surviving Information Overload: A Plea for Balance." *Explorations in Media Ecology* 14, no. 3 (December 2015) 257–74. https://doi. org/10.1386/eme.14.3-4.257_1.

Chun, Wendy Hui Kyong. *Updating to Remain the Same: Habitual New Media*. Kindle ed. Cambridge, MA: MIT Press, 2016.

Clayton, Russell B., Alexander Nagurney, and Jessica R. Smith. "Cheating, Breakup, and Divorce: Is Facebook Use to Blame?" *Cyberpsychology, Behavior, and Social Networking* 16, no. 10 (June 2013) 717–20. https://doi.org/10.1089/ cyber.2012.0424.

Cleveland, Paul A. "Connections Between the Austrian School of Economics and Christian Faith: A Personalist Approach." *Journal of Markets & Morality* 6, no. 2 (2003) 663–73. https://www.marketsandmorality.com/index.php/mandm/article/ view/467.

Clifton, Jacob. "How Did World War II Affect Television?" HowStuffWorks, March 2011. https://people.howstuffworks.com/culture-traditions/tv-and-culture/world-war-ii-affect-television.htm.

Cocking, Dean, Jeroen van den Hoven, and Job Timmermans. "Introduction: One Thousand Friends." *Ethics and Information Technology* 14, no. 3 (January 2012) 179–84. https://doi.org/10.1007/s10676-012-9299-5.

Cocking, Dean, and Steve Matthews. "Unreal Friends." *Ethics and Information Technology* 2, no. 4 (December 2000) 223–31. https://doi.org/10.1023/A:1011414704851.

Cohen, Julie E. *Configuring the Networked Self: Law, Code, and the Play of Everyday Practice*. New Haven, CT: Yale University Press, 2012.

Cooper, Paige. "23 YouTube Statistics That Matter to Marketers in 2020." *Social Media Marketing & Management Dashboard* (blog), December 2019. https://blog. hootsuite.com/youtube-stats-marketers/.

Corey, Elizabeth. "Our Need for Privacy." *First Things* 255 (August 2015) 45–49. https:// www.firstthings.com/article/2015/08/our-need-for-privacy.

Cortright, Joe. "Less in Common." Portland: City Observatory, August 2015. http:// cityobservatory.org/wp-content/files/CityObservatory_Less_In_Common.pdf.

Crawford, Matthew B. "The Cost of Paying Attention." *New York Times*, March 2015. https://www.nytimes.com/2015/03/08/opinion/sunday/the-cost-of-paying-attention.html.

Cringely, Robert. "Welcome to the Post-Decision Age." *I, Cringely* (blog), November 2016. https://www.cringely.com/2016/11/28/post-decision-age/.

Crouch, Andy. *Culture Making: Recovering Our Creative Calling*. Downers Grove, IL: InterVarsity, 2008.

Daniel, Caroline, and Maija Palmer. "Google's Goal: To Organise Your Daily Life." *Financial Times*, May 2007. https://www.ft.com/content/c3e49548-088e-11dc-b11e-000b5df10621.

Darwin, Charles. *The Origin of Species*. New York: P. F. Collier & Son, 1909.

David, Gaby, and Carolina Cambre. "Screened Intimacies: Tinder and the Swipe Logic." *Social Media + Society* 2, no. 2 (April 2016). https://doi. org/10.1177/2056305116641976.

Davies, Craig. "Gaming Industry Boosts Pennsylvania Communities." *CasinoBeats* (blog), May 2019. https://www.casinobeats.com/2019/05/30/gaming-industry-boosts-pennsylvania-communities/.

Davies, John J., Brittany Bird, Casey Chaffin, Joseph Eldridge, Angela Hoover, David Law, Jared Munyan, and Keri Shurtliff. "Habitual, Unregulated Media Use and Marital Satisfaction in Recently Married LDS Couples." *Western Journal of Communication* 76, no. 1 (November 2013) 65–85. https://doi.org/10.1080/105 70314.2012.637541.

Dawson, Christopher. "Religion and Mass Civilization—The Problem of the Future." *The Dublin Review* 213, no. 428 (January 1944) 1–8.

Debes, Remy. "Dignity's Gauntlet." *Philosophical Perspectives* 23 (2009) 45–78. https://www.jstor.org/stable/40658394.

Dennis, Michael Aaron. "Silicon Valley." In *Encyclopedia Britannica*. Chicago: Britannica, September 2019. https://www.britannica.com/place/Silicon-Valley-region-California.

Dong, Guangheng, Yue Shen, Jie Huang, and Xiaoxia Du. "Impaired Error-Monitoring Function in People with Internet Addiction Disorder: An Event-Related FMRI Study." *European Addiction Research* 19, no. 5 (2013) 269–75. https://doi.org/10.1159/000346783.

Doomen, Jasper. "Beyond Dignity." *Archiv Für Begriffsgeschichte* 57 (2015) 57–72. https://www.jstor.org/stable/26353760.

DuckDuckGo. "Measuring the Filter Bubble: How Google Is Influencing What You Click." *DuckDuckGo Blog* (blog), December 2018. https://spreadprivacy.com/google-filter-bubble-study/.

Dufková, Kateřina, Michal Ficek, Lukáš Kencl, Jakub Novak, Jan Kouba, Ivan Gregor, and Jiří Danihelka. "Active GSM Cell-ID Tracking: 'Where Did You Disappear?'" In *Proceedings of the First ACM International Workshop on Mobile Entity Localization and Tracking in GPS-Less Environments*, 7–12. MELT '08. San Francisco: Association for Computing Machinery, 2008. https://doi.org/10.1145/1410012.1410015.

Duke, Éilish, and Christian Montag. "Smartphone Addiction, Daily Interruptions and Self-Reported Productivity." *Addictive Behaviors Reports* 6 (December 2017) 90–95. https://doi.org/10.1016/j.abrep.2017.07.002.

Dunbar, R. I. M. "Do Online Social Media Cut Through the Constraints That Limit the Size of Offline Social Networks?" *Royal Society Open Science* 3, no. 1 (January 2016). https://doi.org/10.1098/rsos.150292.

Dzieza, Josh. "Robots Aren't Taking Our Jobs—They're Becoming Our Bosses." *The Verge* (blog), February 2020. https://www.theverge.com/2020/2/27/21155254/automation-robots-unemployment-jobs-vs-human-google-amazon.

Ekins, Emily. "Poll: 62% of Americans Say They Have Political Views They're Afraid to Share." Washington, DC: Cato Institute, July 2020. https://www.cato.org/publications/survey-reports/poll-62-americans-say-they-have-political-views-theyre-afraid-share.

Ellul, Jacques. *The Technological Society*. Introduction by Robert K. Merton. Translated by John Wilkinson. New York: Vintage, 1964.

Emmons, Libby. "The Transhumanism Revolution: Oppression Disguised as Liberation." *Quillette* (blog), July 2018. https://quillette.com/2018/07/11/the-transhumanism-revolution-oppression-disguised-as-liberation/.

Enderle, Rob. "In the Shadow of Paris: How Watson Could Protect and Nurture Us." *IT Business Edge* (blog), November 2015. https://www.itbusinessedge.com/blogs/

unfiltered-opinion/in-the-shadow-of-paris-how-watson-could-protect-and-nurture-us.html.

Epstein, Richard A. "The Legal Regulation of Genetic Discrimination: Old Responses to New Technology." *Boston University Law Review* 74, no. 1 (1994) 1–23. https://chicagounbound.uchicago.edu/cgi/viewcontent.cgi?article=2314&context=journal_articles.

Epstein, Robert, and Ronald E. Robertson. "The Search Engine Manipulation Effect (SEME) and Its Possible Impact on the Outcomes of Elections." *Proceedings of the National Academy of Sciences* 112, no. 33 (August 2015) E4512–21. https://doi.org/10.1073/pnas.1419828112.

Eswaran, Mukesh, and Hugh M. Neary. "An Economic Theory of the Evolutionary Emergence of Property Rights." *American Economic Journal: Microeconomics* 6, no. 3 (2014) 203–26. https://www.jstor.org/stable/43189681.

Ewen, Stuart. *Captains of Consciousness: Advertising and the Social Roots of the Consumer Culture.* 25th ed. New York: Basic, 2008.

Eyal, Nir. *Hooked: How to Build Habit-Forming Products.* Edited by Ryan Hoover. New York: Portfolio, 2014.

Facebook. "Community Standards." 2020. https://www.facebook.com/communitystandards/.

———. "Data Policy." https://www.facebook.com/about/privacy/your-info.

Farrugia, Rianne. "Facebook and Relationships: A Study of How Social Media Use Is Affecting Long-Term Relationships." MS thesis, Rochester Institute of Technology, 2013. https://www.academia.edu/36070788/Facebook_and_Relationships_A_Study_of_How_Social_Media_Use_is_Affecting_Long-Term_Relationships.

Farseev, Aleksandr, Kirill Lepikhin, Hendrik Schwartz, Eu Khoon Ang, and Kenny Powar. "SoMin.Ai: Social Multimedia Influencer Discovery Marketplace." In *Proceedings of the 26th ACM International Conference on Multimedia*, 1234–236. MM '18. Seoul, Republic of Korea: Association for Computing Machinery, 2018. https://doi.org/10.1145/3240508.3241387.

Feldman, Ronen. "Techniques and Applications for Sentiment Analysis." *Communications of the ACM* 56, no. 4 (April 2013) 82–89. https://doi.org/10.1145/2436256.2436274.

Fernandez, Luke, and Susan J. Matt. *Bored, Lonely, Angry, Stupid: Changing Feelings about Technology, from the Telegraph to Twitter.* Cambridge, MA: Harvard University Press, 2019.

Fioravanti, Giulia, Alfonso Prostamo, and Silvia Casale. "Taking a Short Break from Instagram: The Effects on Subjective Well-Being." *Cyberpsychology, Behavior, and Social Networking* 23, no. 2 (December 2019) 107–12. https://doi.org/10.1089/cyber.2019.0400.

Fogg, B. J. "Foreword." In *Designing for Behavior Change: Applying Psychology and Behavioral Economics.* Kindle ed. Sebastopol, CA: O'Reilly Media, 2013.

Fort, Timothy L., Anjanette Raymond, and Scott Shackelford. "The Angel on Your Shoulder: Prompting Employees to Do the Right Thing Through the Use of Wearables." *Northwestern Journal of Technology and Intellectual Property* 14, no. 2 (2016) 139–70. https://doi.org/10.2139/ssrn.2661069.

Fox, Jesse. "The Dark Side of Social Networking Sites in Romantic Relationships." In *The Psychology of Social Networking: Personal Experience in Online Communities*, 78–88. De Gruyter, 2015. https://www.degruyter.com/view/title/518787.

Frantzman, Seth. "Twitter Censors Trump for Glorifying Violence, Lets Iran Threaten Israel." *The Jerusalem Post* (blog), May 2020. https://www.jpost.com/middle-east/iran-news/twitter-censors-trump-for-glorifying-violence-lets-iran-threaten-israel-629671.

FREE WILL—Lawrence Krauss and Richard Dawkins. 2012. https://www.youtube.com/watch?v=anBxaOcZnGk.

Garfinkel, Simson, John M. Abowd, and Christian Martindale. "Understanding Database Reconstruction Attacks on Public Data." *Communications of the ACM* 62, no. 3 (February 2019) 46–53. https://doi.org/10.1145/3287287.

Garvie, Clare, Alvaro M. Bedoya, Jonathan Frankle, Moriah Daugherty, Katie Evans, Edward J. George, Sabrina McCubbin, et al. "Unregulated Police Face Recognition in America." *Georgetown Law*, October 2016.

Gay, Craig M. *Modern Technology and the Human Future: A Christian Appraisal.* Downers Grove, IL: InterVarsity, 2018.

Geisler, Norman L. "Averroes." In *Baker Encyclopedia of Christian Apologetics*, 63–64. Baker Reference Library. Grand Rapids: Baker, 1999.

Gertz, Nolen. "The Four Facebooks." *The New Atlantis* 58 (Spring 2019) 65–70. https://www.thenewatlantis.com/publications/the-four-facebooks.

———. *Nihilism and Technology.* Kindle ed. Lanham, MD: Rowman & Littlefield, 2018.

Gill, Benjamin. "Megachurch Given the Boot by City, Just Because Pastor 'Liked' Some Conservative Tweets." *CBN News*, June 2020. https://www1.cbn.com/cbnnews/us/2020/june/megachurch-given-the-boot-by-city-just-because-pastor-liked-some-conservative-tweets.

Gillespie, Michael Allen. *The Theological Origins of Modernity.* Chicago: University of Chicago Press, 2008.

Goldhaber, Michael H. "Attention Shoppers!" *Wired*, December 1997. https://www.wired.com/1997/12/es-attention/.

Gomez-Uribe, Carlos A., and Neil Hunt. "The Netflix Recommender System: Algorithms, Business Value, and Innovation." *ACM Transactions on Management Information Systems* 6, no. 4 (January 2016) 1–19. https://doi.org/10.1145/2843948.

Gonzales, Robert, Jr. "Man: God's Visible Replica and Vice-Regent." *Reformed Baptist Theological Review* 5, no. 2 (2008) 63–87.

Gotlieb, Calvin C. "Privacy: A Concept Whose Time Has Come and Gone." In *Computers, Surveillance, and Privacy*, edited by David Lyon and Elia Zureik, 156–74. Minnesota Archive Editions ed. Minneapolis: University of Minnesota Press, 1996.

Grandjean, Martin. "A Social Network Analysis of Twitter: Mapping the Digital Humanities Community." *Cogent Arts & Humanities* 3.1 (2016) 1–14. https://doi.org/10.1080/23311983.2016.1171458.

Grasso, Samantha. "The 10 Best Apps for Shooting and Editing Selfies." *The Daily Dot* (blog), June 2016. https://www.dailydot.com/debug/best-selfie-apps-iphone-android/.

Green, Joel B. *Body, Soul, and Human Life: The Nature of Humanity in the Bible.* Studies in Theological Interpretation. Grand Rapids: Baker Academic, 2008.

Greggo, Stephen P., and Lucas Tillett. "Beyond Bioethics 101: Where Theology Gets Personal and Pastoral." *Journal of the Evangelical Theological Society* 53, no. 2 (2010) 349–64.

Griffiths, Mark. "How Is Technology Innovation Impacting Gambling Addiction?" *National Rehabs Directory* (blog), November 2019. https://www.rehabs.com/pro-talk/technologys-impact-on-gambling-addiction/.

Grosser, Benjamin. "What Do Metrics Want?" *Computational Culture*, no. 4 (November 2014) 1–41. http://computationalculture.net/what-do-metrics-want/.

Grudem, Wayne A. *Systematic Theology: An Introduction to Biblical Doctrine.* Grand Rapids: Zondervan Academic, 2004.

Guillory, Jamie E., and Jeffrey T. Hancock. "Effects of Network Connections on Deception and Halo Effects in LinkedIn." In *The Psychology of Social Networking: Personal Experience in Online Communities*, Vol. 1. De Gruyter, 2015. https://www.degruyter.com/view/title/518787.

Haidt, Jonathan, and Tobias Rose-Stockwell. "The Dark Psychology of Social Networks." *The Atlantic*, December 2019. https://www.theatlantic.com/magazine/archive/2019/12/social-media-democracy/600763/.

Halpern, Mark. "The Trouble with the Turing Test." *The New Atlantis*, no. 11 (2006) 42–63. https://www.jstor.org/stable/43152219.

Hammond, Ron, and Hui-Tzu Grace Chou. "Using Facebook: Good for Friendship but Not so Good for Intimate Relationships." In *The Psychology of Social Networking: Personal Experience in Online Communities*, edited by Giuseppe Riva, Brenda K. Wiederhold, Pietro Cipresso, and Aneta Przepiórka, 41–52. Warsaw: De Gruyter Open, 2016.

Hasse, Dag Nikolaus. "Influence of Arabic and Islamic Philosophy on the Latin West." *Stanford Encyclopedia of Philosophy*, September 2008. https://plato.stanford.edu/entries/arabic-islamic-influence/.

Hawley, Josh. "The Big Tech Threat." *First Things* (blog), May 2019. https://www.firstthings.com/web-exclusives/2019/05/the-big-tech-threat.

Hayes, Andrew F., Carroll J. Glynn, and James Shanahan. "Validating the Willingness to Self-Censor Scale: Individual Differences in the Effect of the Climate of Opinion on Opinion Expression." *International Journal of Public Opinion* 17, no. 4 (March 2005). https://www.academia.edu/280701/Validating_the_Willingness_to_Self-Censor_Scale_Individual_Differences_In_the_Effect_of_the_Climate_of_Opinion_on_Opinion_Expression.

Heilbroner, Robert. "Technological Determinism Revisited." In *Does Technology Drive History? The Dilemma of Technological Determinism*, edited by Merritt Roe Smith and Leo Marx, 67–78. Cambridge, MA: The MIT Press, 1994.

Helmer, Sven. "May I Have Your Attention, Please: Building a Dystopian Attention Economy." In *Companion of the Web Conference 2018*, 1529–33. Association of Computing Machinery, 2018. https://doi.org/10.1145/3184558.3191605.

Henderson, Conor M., Joshua T. Beck, and Robert W. Palmatier. "Review of the Theoretical Underpinnings of Loyalty Programs." *Journal of Consumer Psychology* 21, no. 3 (2011) 256–76. https://www.jstor.org/stable/23049324.

Henry, Carl F. H. *God, Revelation, and Authority.* 6 vols. Wheaton, IL: Crossway, 1999.

Henry, Leslie Meltzer. "The Jurisprudence of Dignity." *University of Pennsylvania Law Review* 160, no. 1 (2011) 169–233. https://www.jstor.org/stable/41308486.

Highfield, Ron. *God, Freedom and Human Dignity: Embracing a God-Centered Identity in a Me-Centered Culture.* Kindle ed. Downers Grove, IL: InterVarsity, 2013.

Hoekema, Anthony A. *Created in God's Image.* Grand Rapids: Eerdmans, 1994.

Hořeňovský, Martin. "Modern SAT Solvers: Fast, Neat and Underused (Part 1 of N)."
    The Coding Nest (blog), August 2018. http://codingnest.com/modern-sat-solvers-
    fast-neat-underused-part-1-of-n/.
Horne, Charles M. "Christian Humanism." *Journal of the Evangelical Theological Society*
    14, no. 3 (1971) 185–191.
Hosie, Rachel. "People Want to Look Like Versions of Themselves with Filters Rather
    Than Celebrities, Cosmetic Doctor Says." *The Independent* (blog), February 2018.
    http://www.independent.co.uk/life-style/cosmetic-surgery-snapchat-instagram-
    filters-demand-celebrities-doctor-dr-esho-london-a8197001.html.
Hua, Jingyu, Zhenyu Shen, and Sheng Zhong. "We Can Track You If You Take the Metro."
    IEEE Transactions on Information Forensics and Security 12, no. 2 (February 2017)
    286–97. https://doi.org/10.1109/TIFS.2016.2611489.
Hughes, Thomas P. "Technological Momentum." In *Does Technology Drive History?*
    The Dilemma of Technological Determinism, edited by Merritt Roe Smith and Leo
    Marx, 101–14. Cambridge, MA: The MIT Press, 1994.
Hunt, Melissa G., Rachel Marx, Courtney Lipson, and Jordyn Young. "No More
    FOMO: Limiting Social Media Decreases Loneliness and Depression." *Journal of
    Social and Clinical Psychology* 37, no. 10 (December 2018) 751–68. https://doi.
    org/10.1521/jscp.2018.37.10.751.
Huston, Geoff. "The Death of Transit?" *APNIC* (blog), January 2018. https://blog.apnic.
    net/2016/10/28/the-death-of-transit/.
"Internet Security Threat Report." Symantec, March 2018. https://www.symantec.com/
    content/dam/symantec/docs/reports/istr-23-2018-en.pdf.
Jacobi, Jennifer A, Eric A Benson, and Gregory D Linden. Personalized
    Recommendations of Item's Represented Within a Database. 7,113,917 B2. Seattle,
    WA, n.d.
Jacobs, Alan. "Attending to Technology: Theses for Disputation." *The New Atlantis*,
    (Winter 2016) 16–45. https://www.thenewatlantis.com/publications/attending-
    to-technology-theses-for-disputation.
Jaffe, Eric. "Toronto Tomorrow." Sidewalk Labs, 2019. https://storage.googleapis.com/
    sidewalk-toronto-ca/wp-content/uploads/2019/06/23135500/MIDP_Volume0.
    pdf.
Jagannathan, Anand. "Frictionless Sharing: Realizing the Promise of Real-Time
    Serendipity." *Engage.Social* (blog), May 2017. https://engage.social/blog/social-
    share/frictionless-sharing-realizing-the-promise-of-real-time-serendipity/.
Jashinsky, Emily. "The Tail of Twitter Is Wagging the Dog of Media." *The Federalist*
    (blog), February 2019. http://thefederalist.com/2019/02/08/the-tail-of-twitter-is-
    wagging-the-dog-of-media/.
Jenkins, Holman W., Jr. "Google and the Search for the Future." *Wall Street Journal*,
    August 2010. https://www.wsj.com/articles/SB10001424052748704901104575423
    294099527212.
Jercinovic, Jason. "Markets of One: After Years of Hyperbole, Mass Customization Is
    Finally Within Our Grasp." *IBM IX* (blog), November 2016. https://www.ibm.
    com/blogs/insights-on-business/ibmix/markets-of-one/.
Jordan, John M. *Machine-Age Ideology: Social Engineering and American Liberalism,
    1911–1939*. Kindle ed. Chapel Hill, NC: University of North Carolina Press, 2010.
Josephonson, Colin, and Yan Shvartzshnaider. "Every Move You Make, I'll Be Watching
    You: Privacy Implications of the Apple U1 Chip and Ultra-Wideband." *Freedom*

to Tinker (blog), December 2019. https://freedom-to-tinker.com/2019/12/21/every-move-you-make-ill-be-watching-you-privacy-implications-of-the-apple-u1-chip-and-ultra-wideband/.

Kahneman, Daniel. *Thinking, Fast and Slow*. New York: Farrar, Straus and Giroux, 2011.

Kaldestad, Øyvind H. "Out of Control: A Review of Data Sharing by Popular Mobile Apps" and "Out of Control: How Consumers Are Exploited by the Online Advertising Industry." Norwegian Consumer Council, January 2020. https://www.forbrukerradet.no/undersokelse/no-undersokelsekategori/report-out-of-control/.

Kant, Immanuel, and J. H. von Kirchmann. *Grundlegung zur metaphysik der sitten*. Portsmith, NH: Heimann, 1870.

Kanter, Jeffrey Andrew, Mitu Singh, and Daniel Gregory Muriello. Moderating content in an online forum. USPA 10356024, filed November 30, 2011, and issued July 16, 2019. http://patft1.uspto.gov/netacgi/nph-Parser?patentnumber=10356024.

Kay, Jonathan. "The Tyranny of Twitter: How Mob Censure Is Changing the Intellectual Landscape." *National Post* (blog), July 2017. https://nationalpost.com/news/world/jonathan-kay-on-the-tyranny-of-twitter-how-mob-censure-is-changing-the-intellectual-landscape.

Kendall, R. T. *Understanding Theology*. Logos Electronic. 3 vols. Scotland: Christian Focus, 2000.

King, Gary, Jennifer Pan, and Margaret Roberts. "How Censorship in China Allows Government Criticism but Silences Collective Expression." *American Political Science Review* 107, no. 2 (May 2013) 1–18.

Kirkpatrick, Keith. "Tracking Shoppers." *Communications of the ACM* 63, no. 2 (January 22, 2020) 19–21. https://doi.org/10.1145/3374876.

Kissling, Paul J. *Genesis*. 2 vols. The College Press NIV Commentary. Joplin, MO: College, 2004.

Klaus, Mika. "Privacy Statements; Difficult to Read and to Understand." *SecJure* (blog), October 22, 2015. http://www.secjure.nl/2015/10/22/privacy-statements-difficult-to-read-and-to-understand/.

Knight, Jon. "6 Reasons You Might Actually Want to Give Google Your Location Data." *Gadget Hacks* (blog), March 2018. https://smartphones.gadgethacks.com/how-to/6-reasons-you-might-actually-want-give-google-your-location-data-0183740/.

Kofman, Ava. "Google's Sidewalk Labs Plans to Package and Sell Location Data on Millions of Cellphones." *The Intercept* (blog), January 2019. https://theintercept.com/2019/01/28/google-alphabet-sidewalk-labs-replica-cellphone-data/.

Krauss, Lawrence M. *A Universe from Nothing: Why There Is Something Rather Than Nothing*. New York: Free, 2012.

Krebs, Dennis L. "Morality: An Evolutionary Account." *Perspectives on Psychological Science* 3, no. 3 (2008) 149–72. https://www.jstor.org/stable/40212242.

Kristof, Kathy. "How Amazon Uses 'Surge Pricing,' Just Like Uber." *CBS News*, July 2017. https://www.cbsnews.com/news/amazon-surge-pricing-are-you-getting-ripped-off-small-business/.

Kugler, Logan. "Being Recognized Everywhere." *Communications of the ACM* 62, no. 2 (January 2019) 17–19. https://doi.org/10.1145/3297803.

———. "The War Over the Value of Personal Data." *Communications of the ACM* 61, no. 2 (January 23, 2018) 17–19. https://doi.org/10.1145/3171580.

Kuss, Daria J., and Mark D. Griffiths. "Social Networking Sites and Addiction: Ten Lessons Learned." *International Journal of Environmental Research and Public Health* 14, no. 3 (March 2017). https://doi.org/10.3390/ijerph14030311.

Leetaru, Kalev. "How Data Brokers and Pharmacies Commercialize Our Medical Data." *Forbes*, April 2018. https://www.forbes.com/sites/kalevleetaru/2018/04/02/how-data-brokers-and-pharmacies-commercialize-our-medical-data/.

———. "How Social Media Is Teaching Us to Emphasize the What Over the Why." *Forbes*, June 2018. https://www.forbes.com/sites/kalevleetaru/2018/06/16/how-social-media-is-teaching-us-to-emphasize-the-what-over-the-why/.

Leupold, H. C. *Exposition of the Psalms.* Grand Rapids: Baker, 1977.

"Levels of Problem Gambling in England." https://www.gamblingcommission.gov.uk/news-action-and-statistics/Statistics-and-research/Levels-of-participation-and-problem-gambling/Levels-of-problem-gambling-in-England.aspx.

Levy, Adam. "How LinkedIn Earns a Higher Gross Profit Margin Than Facebook and Google." *The Motley Fool*, March 2015. https://www.fool.com/investing/general/2015/03/11/how-linkedin-earns-a-higher-gross-profit-margin-th.aspx.

Lewis, C. S. *The Abolition of Man.* New York: HarperCollins, 2009.

———. *Miracles.* New York: HarperCollins, 2009.

———. *Surprised by Joy: The Shape of My Early Life.* New York: Mariner, 1966.

Lewis, Dan. "Thomas Edison Drove the Film Industry to California." *Mental Floss*, July 2013. https://www.mentalfloss.com/article/51722/thomas-edison-drove-film-industry-california.

Lewis, Gordon R., and Bruce A. Demarest. *Integrative Theology: Knowing Ultimate Reality: The Living God.* Vol. 1. Grand Rapids: Zondervan, 1987.

Light, Jennifer S. "When Computers Were Women." *Technology and Culture* 40, no. 3 (1999) 455–83. https://www.jstor.org/stable/25147356.

Linné, Carl von, Johann Friedrich Gmelin, Robert Kerr, J. Archer, and G. Kinnear. *The Animal Kingdom.* London: A. Strahan, 1792.

Lints, Richard. *Identity and Idolatry: The Image of God and Its Inversion.* Edited by D. A. Carson. Vol. 36. New Studies in Biblical Theology. Leicester: Apollos, 2015.

Lombardi, Claudio. "The Illusion of a 'Marketplace of Ideas' and the Right to Truth." *American Affairs Journal* 3, no. 1 (Spring 2019) 14. https://americanaffairsjournal.org/2019/02/the-illusion-of-a-marketplace-of-ideas-and-the-right-to-truth/.

Lovink, Geert. *Sad by Design: On Platform Nihilism.* Kindle ed. London: Pluto, 2019.

Lynch, Clifford. "The Rise of Reading Analytics and the Emerging Calculus of Reader Privacy in the Digital World." *First Monday* 22, no. 4 (2017) 1–22. https://doi.org/10.5210/fm.v22i4.7414.

MacKenzie, Ian, Chris Meyer, and Steve Noble. "How Retailers Can Keep Up with Consumers." *McKinsey* (blog), October 2013. https://www.mckinsey.com/industries/retail/our-insights/how-retailers-can-keep-up-with-consumers.

Manjoo, Farhad. "Facebook's Terrible Plan To Make Us Share Everything We Do Online." *Slate Magazine* (blog), September 2011. https://slate.com/technology/2011/09/facebook-s-terrible-plan-to-make-us-share-everything-we-do-online.html.

Manson, Mark. "In the Future, Our Attention Will Be Sold." *Mark Manson* (blog), December 2014. https://markmanson.net/attention.

Marx, Jesse. "The Mission Creep of Smart Streetlights." *Voice of San Diego* (blog), February 2020. https://www.voiceofsandiego.org/topics/public-safety/the-mission-creep-of-smart-streetlights/.

Marx, Karl. *The Poverty of Philosophy*. Moscow: Foreign Language Publishing House, 1892.

Mateescu, Alexandra, and Aiha Nguyen. "Explainer: Algorithmic Management in the Workplace." Data & Society, February 2019. https://datasociety.net/output/explainer-algorithmic-management-in-the-workplace/.

———. "Explainer: Workplace Monitoring & Surveillance." Data & Society, February 2019. https://datasociety.net/output/explainer-workplace-monitoring-surveillance/.

Mathews, K. A. *Genesis 1–11:26*. Vol. 1A. The New American Commentary. Nashville: Broadman & Holman, 1996.

Mathur, Arunesh, Gunes Acar, Michael Friedman, Elena Lucherini, Jonathan Mayer, Marshini Chetty, and Arvind Narayanan. "Dark Patterns at Scale: Findings from a Crawl of 11K Shopping Websites." In *Proceedings of the ACM on Human-Computer Interaction*, 3:81:1–81:32. Association of Computing Machinery, 2019. http://doi.acm.org/10.1145/3359183.

May, Gerald G. *Addiction and Grace: Love and Spirituality in the Healing of Addictions*. Reissue ed. New York: HarperOne, 2007.

McCarthy, Caroline. "Silicon Valley Has a Problem with Conservatives. But Not the Political Kind." *Vox*, June 2018. https://www.vox.com/first-person/2018/6/12/17443134/silicon-valley-conservatives-religion-atheism-james-damore.

McDermott, Rose, James Fowler, and Nicholas Christakis. "Breaking Up Is Hard to Do, Unless Everyone Else Is Doing It Too: Social Network Effects on Divorce in a Longitudinal Sample." *Social Forces; a Scientific Medium of Social Study and Interpretation* 92, no. 2 (December 2013) 491–519. https://doi.org/10.1093/sf/sot096.

McGregor, Shannon C., and Logan Molyneux. "Twitter's Influence on News Judgment: An Experiment Among Journalists." *Journalism*, October 2018, 1–17. https://doi.org/10.1177/1464884918802975.

McInerny, Ralph, and Saint Thomas (Aquinas). *Aquinas Against the Averroists: On There Being Only One Intellect*. West Lafayette, IN: Purdue University Press, 1993.

McLean, Siân A., Susan J. Paxton, Eleanor H. Wertheim, and Jennifer Masters. "Photoshopping the Selfie: Self Photo Editing and Photo Investment Are Associated with Body Dissatisfaction in Adolescent Girls." *International Journal of Eating Disorders* 48, no. 8 (2015) 1132–40. https://doi.org/10.1002/eat.22449.

McLuhan, Marshall. *The Essential McLuhan*. Edited by Frank Zingrone and Marshall McLuhan. New York: Basic, 1996.

McNutt, Paula M. "The Kenites, the Midianites, and the Rechabites as Marginal Mediators in Ancient Israelite Tradition." *Semeia* 67 (1994) 109–33.

McPherson, Miller, Lynn Smith-Lovin, and James M. Cook. "Birds of a Feather: Homophily in Social Networks." *Annual Review of Sociology* 27.1 (2001) 415–44. https://doi.org/10.1146/annurev.soc.27.1.415.

Meadowcroft, Micah. "The Distance Between Us." *The New Atlantis* 58 (Spring 2019) 74–79. https://www.thenewatlantis.com/publications/the-distance-between-us.

Meeker, Mary. "Internet Trends 2019." http://bondcap.com/report/it19.

Melville, George W. "The Engineer's Duty as a Citizen." *Transactions of the American Society of Mechanical Engineers* 32 (1910) 527–32.

Metz, Rachel. "Yes, Alexa Is Recording Mundane Details of Your Life, and It's Creepy as Hell." *MIT Technology Review* (blog), May 2018. https://www.technologyreview.com/s/611216/yes-alexa-is-recording-mundane-details-of-your-life-and-its-creepy-as-hell/.

Meyer, Michael J. "Dignity, Rights, and Self-Control." *Ethics* 99, no. 3 (1989) 520–34. https://www.jstor.org/stable/2380864.

Middleton, J. Richard. *The Liberating Image: The Imago Dei in Genesis 1.* Grand Rapids: Brazos, 2005.

Miłkowski, Marcin. "Why Think That the Brain Is Not a Computer?" *APA Newsletter of Philosophy and Computers* 16, no. 2 (Spring 2016) 22–28. https://philpapers.org/rec/MIKWTT.

Miller, Jayne. "This Is What Our 2019 Submarine Cable Map Shows Us About Content Provider Cables." *TeleGeography* (blog), March 2019. https://blog.telegeography.com/this-is-what-our-2019-submarine-cable-map-shows-us-about-content-provider-cables.

Mills, Jennifer S., Sarah Musto, Lindsay Williams, and Marika Tiggemann. "'Selfie' Harm: Effects on Mood and Body Image in Young Women." *Body Image* 27 (December 2018) 86–92. https://doi.org/10.1016/j.bodyim.2018.08.007.

MIT Technology Review Insights. "The Rise of Data Capital." *MIT Technology Review*, March 2016. https://www.technologyreview.com/s/601081/the-rise-of-data-capital/.

Moby and The Void Pacific Choir. "Are You Lost In The World Like Me?" https://youtu.be/VASywEuqFd8.

Mohajeri Moghaddam, Hooman, Gunes Acar, Ben Burgess, Arunesh Mathur, Danny Yuxing Huang, Nick Feamster, Edward W. Felten, Prateek Mittal, and Arvind Narayanan. "Watching You Watch: The Tracking Ecosystem of Over-the-Top TV Streaming Devices." In *Proceedings of the 2019 ACM SIGSAC Conference on Computer and Communications Security—CCS '19*, 131–47. Association of Computing Machinery, 2019. https://doi.org/10.1145/3319535.3354198.

Moore, Adam D. "Privacy, Speech, and Values: What We Have No Business Knowing." *Ethics and Information Technology* 18, no. 1 (January 2016) 41–49. https://doi.org/10.1007/s10676-016-9397-x.

Moretta, John Anthony. *The Hippies: A 1960s History.* Jefferson, NC: McFarland, 2017.

Mull, Amanda. "Gadgets for Life on a Miserable Planet." *The Atlantic* (blog), January 2020. https://www.theatlantic.com/health/archive/2020/01/why-do-people-still-love-consumer-tech/604909/.

Mumford, Lewis. *Technics and Civilization.* London: George Routledge & Sons, 1947.

Munn, Nicholas John. "The Reality of Friendship Within Immersive Virtual Worlds." *Ethics and Information Technology* 14, no. 1 (March 2012) 1–10. https://doi.org/10.1007/s10676-011-9274-6.

Murrill, Brandon J., and Edward C. Liu. "Smart Meter Data: Privacy and Cybersecurity." Congressional Research Service, February 2012. https://fas.org/sgp/crs/misc/R42338.pdf.

Nair, Hari. "A Predictive Analytics Model to Determine the Flight Risk Score of Employees." *LinkedIn* (blog), January 2019. https://www.linkedin.com/pulse/predictive-analytics-model-determine-flight-risk-score-dr-hari-nair.

"The National Economic Benefits of the Canadian Gaming Industry: Key Findings Report." Canadian Gaming Association, 2017. http://canadiangaming.ca/wp-content/uploads/CGA_KeyFindings_document_D.pdf.

Newlands, Gemma, Christoph Lutz, and Christian Fieseler. "The Conditioning Function of Rating Mechanisms for Consumers in the Sharing Economy." *Internet Research* 29, no. 5 (January 2019) 1090–1108. https://doi.org/10.1108/INTR-03-2018-0134.

Ninivaggi, Frank J. "Loneliness: A New Epidemic in the USA." *Psychology Today* (blog), February 2019. https://www.psychologytoday.com/blog/envy/201902/loneliness-new-epidemic-in-the-usa.

Noell, Edd. "An End to Scarcity? Keynes's Moral Critique of Capitalism and Its Ambiguous Legacy." *Journal of Markets & Morality* 20, no. 1 (2017) 39–50.

Noelle-Neumann, Elisabeth. *The Spiral of Silence: Public Opinion—Our Social Skin.* 2d ed. Chicago: University of Chicago Press, 1993.

Novak, Matt. "Predictions for Educational TV in the 1930s." *Smithsonian* (blog), May 2012. https://www.smithsonianmag.com/history/predictions-for-educational-tv-in-the-1930s-107574983/.

Novak, Michael. "The Judeo-Christian Foundation of Human Dignity, Personal Liberty, and the Concept of the Person." *Journal of Markets & Morality* 1, no. 2 (1998) 107–21. https://www.marketsandmorality.com/index.php/mandm/article/view/2.

Noyes, Dan. "Top 20 Facebook Statistics—Updated August 2020." *Zephoria Inc.* (blog), August 2020. https://zephoria.com/top-15-valuable-facebook-statistics/.

O'Brien, Danny. "Massive Database Leak Gives Us a Window into China's Digital Surveillance State." *Electronic Frontier Foundation* (blog), March 2019. https://www.eff.org/deeplinks/2019/03/massive-database-leak-gives-us-window-chinas-digital-surveillance-state.

Orben, Amy, and Andrew K. Przybylski. "The Association Between Adolescent Well-Being and Digital Technology Use." *Nature Human Behaviour* 3 (January 2019) 173–82. https://doi.org/10.1038/s41562-018-0506-1.

Osburn, Madeline. "How Chrissy Teigen And The New York Times Cancelled Alison Roman." *The Federalist* (blog), May 2020. https://thefederalist.com/2020/05/27/how-chrissy-teigen-and-the-new-york-times-canceled-alison-roman/.

Osman, Maddy. "Mind-Blowing LinkedIn Statistics and Facts (2020)." *Kinsta Managed WordPress Hosting* (blog), April 2020. https://kinsta.com/blog/linkedin-statistics/.

Packard, Vance, and Mark Crispin Miller. *The Hidden Persuaders.* Brooklyn, NY: Ig, 2007.

Pardes, Arielle. "How Facebook and Other Sites Manipulate Your Privacy Choices." *Wired,* August 2020. https://www.wired.com/story/facebook-social-media-privacy-dark-patterns/.

Pariser, Eli. *The Filter Bubble: How the New Personalized Web Is Changing What We Read and How We Think.* New York: Penguin, 2011.

Patrick, Dale. "Studying Biblical Law as a Humanities." *Semeia* 45 (1988) 27–48.

Paul, Kari. "How Rating Everything from Your Uber Driver to Your Airbnb Host Has Become a Nightmare." *MarketWatch* (blog), May 2019. https://www.marketwatch.com/story/how-rating-everything-from-your-uber-driver-to-your-airbnb-host-has-become-a-nightmare-2019-04-01.

Pearcey, Nancy, and Charles Thaxton. *The Soul of Science: Christian Faith and Natural Philosophy.* Irvine, CA: Good News, 1994.

Penney, Jonathon. "Chilling Effects: Online Surveillance and Wikipedia Use." *Berkeley Technology Law Journal* 31, no. 1 (January 2016) 117–75. https://doi.org/10.15779/Z38SS13.

Perez, Beatrice, Mirco Musolesi, and Gianluca Stringhini. "You Are Your Metadata." *ArXiv*, 2018, 1–10. https://arxiv.org/abs/1803.10133.

Pinckaers, Servais. *The Sources of Christian Ethics*. Translated by Mary Thomas Noble. Washington, DC: Catholic University of America Press, 1995.

Justia Law. *Planned Parenthood of Southeastern Pa. v. Casey*, 505 U.S. 833 (1992). https://supreme.justia.com/cases/federal/us/505/833/.

Plato. *The Dialogues of Plato*. Translated by B. Jowett. 3rd ed. Oxford: Clarendon, 1892.

Pomerantsev, Peter. "Beyond Propaganda." *Foreign Policy* (blog), June 2015. https://foreignpolicy.com/2015/06/23/beyond-propaganda-legatum-transitions-forum-russia-china-venezuela-syria/.

————. "Disinformation All the Way Down." *The American Interest* (blog), February 2018. https://www.the-american-interest.com/2018/02/27/disinformation-all-the-way-down/.

"Port Huron Statement." 1962. http://www.sds-1960s.org/PortHuronStatement-draft.htm.

Postman, Neil, and Andrew Postman. *Amusing Ourselves to Death: Public Discourse in the Age of Show Business*. Harmondsworth: Penguin, 2005.

Powers, William. *Hamlet's BlackBerry: A Practical Philosophy for Building a Good Life in the Digital Age*. Kindle ed. New York: Harper Perennial, 2010.

Pretrial Justice Institute. "Risk Assessment: Evidence-Based Pretrial Decision-Making." Pretrial Justice Institute, August 2016. https://university.pretrial.org/viewdocument/risk-assessment-evid.

Primack, Brian A., Ariel Shensa, Jaime E. Sidani, Erin O. Whaite, Liu yi Lin, Daniel Rosen, Jason B. Colditz, Ana Radovic, and Elizabeth Miller. "Social Media Use and Perceived Social Isolation Among Young Adults in the U.S." *American Journal of Preventive Medicine* 53, no. 1 (July 2017) 1–8. https://doi.org/10.1016/j.amepre.2017.01.010.

Rajanala, Susruthi, Mayra B. C. Maymone, and Neelam A. Vashi. "Selfies—Living in the Era of Filtered Photographs." *JAMA Facial Plastic Surgery* 20, no. 6 (December 2018) 443–44. https://doi.org/10.1001/jamafacial.2018.0486.

Raphael, Rina. "Netflix CEO Reed Hastings: Sleep Is Our Competition." *Fast Company* (blog), November 2017. https://www.fastcompany.com/40491939/netflix-ceo-reed-hastings-sleep-is-our-competition.

Razaghpanah, Abbas, Rishab Nithyanand, Narseo Vallina-Rodriguez, Srikanth Sundaresan, Mark Allman, Christian Kreibich, and Phillipa Gill. "Apps, Trackers, Privacy, and Regulators: A Global Study of the Mobile Tracking Ecosystem." In *The Network and Distributed System Security Symposium 2018*, 1–15. http://eprints.networks.imdea.org/1744/.

Reardon, Joel, Álvaro Feal, and Primal Wijesekera. "50 Ways to Leak Your Data." In *28th USNIX Security Symposium*, 603–20. USENIX, 2019. https://www.usenix.org/conference/usenixsecurity19/presentation/reardon.

"Recommendations for Social Media Use on the National Desk." *Washington Post*, April 2020. https://int.nyt.com/data/documenthelper/7010-recommendations-for-social-med/a5c91e59333f4faoc8bf/optimized/full.pdf#page=1.

Rejoiner. "Amazon's Recommendation Engine: The Secret to Selling More Online." 2015. http://rejoiner.com/resources/amazon-recommendations-secret-selling-online/.

Ren, Jingjing, Daniel J. Dubois, David Choffnes, Anna Maria Mandalari, Roman Kolcun, and Hamed Haddadi. "Information Exposure from Consumer IOT Devices: A Multidimensional, Network-Informed Measurement Approach." In *Proceedings of the Internet Measurement Conference on—IMC '19*, 267–79. Association of Computing Machinery, 2019. https://doi.org/10.1145/3355369.3355577.

Richardson, Rashida, Jason Schultz, and Kate Crawford. "Dirty Data, Bad Predictions: How Civil Rights Violations Impact Police Data, Predictive Policing Systems, and Justice." *New York Law Review Online* 94, no. 192 (October 2019) 1–30. https://papers.ssrn.com/sol3/papers.cfm?abstract_id=3333423##.

Rideout, Victoria, and Michael Robb. "Social Media, Social Life." Survey, Common Sense Media, 2018. https://www.commonsensemedia.org/sites/default/files/uploads/research/2018_cs_socialmediasociallife_fullreport-final-release_2_lowres.pdf.

Rifkin, Jeremy. *The Age of Access: The New Culture of Hypercapitalism*. Los Angeles: TarcherPerigee, 2001.

Riley, James Patrick. "'The Frightening Power of Social Media Bans." *American Greatness*, March 2019. https://amgreatness.com/2019/03/14/the-frightening-power-of-social-media-bans/.

Rinehart, Will. "Hearing on Data Ownership: Exploring Implications for Data Privacy Rights and Data Valuation." Presented at the Committee on Banking, Housing, and Urban Affairs, United States Senate, October 2019. https://www.americanactionforum.org/testimony/hearing-on-data-ownership-exploring-implications-for-data-privacy-rights-and-data-valuation/.

Rosen, Larry D., Nancy Cheever, and L. Mark Carrier, eds. *The Wiley Handbook of Psychology, Technology, and Society*. Malden, MA: Wiley-Blackwell, 2015.

Rotondi, Valentina, Luca Stanca, and Miriam Tomasuolo. "Connecting Alone: Smartphone Use, Quality of Social Interactions and Well-Being." *Journal of Economic Psychology* 63 (December 2017) 17–26. https://doi.org/10.1016/j.joep.2017.09.001.

Rousseau, Jean-Jacques. *The Social Contract & Discourses*. Edited by Ernest Rhys. Translated by G. D. H. Cole. Everyman's Library: Philosophy and Theology. New York: Dutton, 1913.

Rushkoff, Douglas. "The Anti-Human Religion of Silicon Valley." *Medium* (blog), December 2018. https://medium.com/s/douglas-rushkoff/the-anti-human-religion-of-silicon-valley-ac37d5528683.

Rutledge, Pamela. "Social Media Addiction: Engage Brain Before Believing." *Psychology Today* (blog), May 2010. http://www.psychologytoday.com/blog/positively-media/201005/social-media-addiction-engage-brain-believing.

Ryrie, Charles Caldwell. *Basic Theology: A Popular Systematic Guide to Understanding Biblical Truth*. Chicago: Moody, 1999.

Sagan, Carl. *Cosmos*. New York: Ballantine, 2013.

Salzer, Samuel, and Silja Voomla. "Behavioral Design 2020 and Beyond." March 2020. https://medium.com/behavior-design-hub/behavioral-design-2020-and-beyond-dc88a87f3b97.

Samson, Alain. "Big Data Is Nudging You." *Psychology Today*, August 2016. https://www.psychologytoday.com/blog/consumed/201608/big-data-is-nudging-you.

Sarna, Nahum M. *Genesis*. Philadelphia: Jewish Publication Society of America, 1989.

Sax, Marijn. "Big Data: Finders Keepers, Losers Weepers?" *Ethics and Information Technology* 18, no. 1 (March 2016) 25–31. https://doi.org/10.1007/s10676-016-9394-0.

Schaef, Anne Wilson. *When Society Becomes an Addict*. Princeton, NJ: HarperOne, 1988.

Schall, Daniel. *Social Network-Based Recommender Systems*. Kindle ed. New York: Springer, 2015.

Schneier, Bruce. *Data and Goliath: The Hidden Battles to Collect Your Data and Control Your World*. New York: Norton, 2016.

Schüll, Natasha Dow. *Addiction by Design: Machine Gambling in Las Vegas*. Princeton, NJ: Princeton University Press, 2014.

Schwartz, Adam. "Chicago's Video Surveillance Cameras." *Northwestern Journal of Technology and Intellectual Property* 11, no. 2 (2013) 47–60. https://scholarlycommons.law.northwestern.edu/njtip/vol11/iss2/4.

Seymour, Richard. *The Twittering Machine*. London: Indigo, 2019.

Shearer, Elisa, and Jeffrey Gottfried. "News Use Across Social Media Platforms 2017." Pew Research Center, August 2017. https://www.journalism.org/2017/09/07/news-use-across-social-media-platforms-2017/.

Shermer, Michael. "How the Survivor Bias Distorts Reality." *Scientific American* (blog), September 2014. https://doi.org/10.1038/scientificamerican0914-94.

Singh, Devin. "Review of *The Origins of Neoliberalism: Modeling the Economy from Jesus to Foucault* by Dotan Leshem." *Syndicate* 4, no. 2 (2017) 58–60.

Skinner, B. F. *Beyond Freedom and Dignity*. New York: Knopf, 1971.

Smith, Eric Alden. "Agency and Adaptation: New Directions in Evolutionary Anthropology." *Annual Review of Anthropology* 42 (2013) 103–20. https://www.jstor.org/stable/43049293.

Smith, Merritt Roe. "Technological Determinism in American Culture." In *Does Technology Drive History? The Dilemma of Technological Determinism*, edited by Merritt Roe Smith and Leo Marx, 1–36. Cambridge, MA: The MIT Press, 1994.

Smith, Merritt Roe, and Leo Marx. "Introduction." In *Does Technology Drive History? The Dilemma of Technological Determinism,* edited by Merritt Roe Smith and Leo Marx, ix–xv. Cambridge, MA: The MIT Press, 1994.

Solove, Daniel J. *Understanding Privacy*. Kindle ed. Cambridge, MA: Harvard University Press, 2008.

Søraker, Johnny Hartz. "How Shall I Compare Thee? Comparing the Prudential Value of Actual Virtual Friendship." *Ethics and Information Technology* 14, no. 3 (September 2012) 209–19. https://doi.org/10.1007/s10676-012-9294-x.

Spiegel, Amy Rose, and Broadly Staff. "I Got Surgery to Look Like My Selfie Filters." *Broadly* (blog), December 2018. https://broadly.vice.com/en_us/article/mby5by/cosmetic-plastic-surgery-social-media-seflies.

Starcevic, Vladan, and Elias Aboujaoude. "Internet Addiction: Reappraisal of an Increasingly Inadequate Concept." *CNS Spectrums* 22, no. 1 (February 2017) 7–13. https://doi.org/10.1017/S1092852915000863.

Steers, Mai-Ly N., Robert E. Wickham, and Linda K. Acitelli. "Seeing Everyone Else's Highlight Reels." *Journal of Social and Clinical Psychology* 33, no. 8 (October 2014) 701–31. https://doi.org/10.1521/jscp.2014.33.8.701.

Steinberg, Brooke. "Disney Fans Say Splash Mountain Needs Update Due to Racism." *New York Post* (blog), June 2020. https://nypost.com/2020/06/10/disney-fans-say-splash-mountain-should-be-re-themed-due-to-racism/.

Stent, G. S. *The Coming of the Golden Age.* Garden City, NY: Natural History, 1969.

Stoddart, Eric. *Theological Perspectives on a Surveillance Society: Watching and Being Watched.* Burlington, VT: Routledge, 2011.

Strong, Augustus Hopkins. *Systematic Theology.* Philadelphia: Griffith & Rowland, 1907.

Summers, Christopher A., Robert W. Smith, and Rebecca Walker Reczek. "An Audience of One: Behaviorally Targeted Ads as Implied Social Labels." *Journal of Consumer Research* 43, no. 1 (June 2016) 156–78. https://doi.org/10.1093/jcr/ucw012.

Tajitsu, Naomi. "Toyota, Panasonic to Set up Company for 'Connected' Homes." *One America News* (blog), May 2019. https://www.oann.com/toyota-panasonic-to-set-up-firm-to-connect-cars-homes-kyodo/.

Tams, Stefan, Renaud Legoux, and Pierre-Majorique Léger. "Smartphone Withdrawal Creates Stress: A Moderated Mediation Model of Nomophobia, Social Threat, and Phone Withdrawal Context." *Computers in Human Behavior* 81 (April 2018) 1–9. https://doi.org/10.1016/j.chb.2017.11.026.

Thaler, Richard H., and Cass R. Sunstein. *Nudge: Improving Decisions About Health, Wealth, and Happiness.* Harmondsworth: Penguin, 2009.

Thomas, Emily. "Cyber-Stalking: When Looking at Other People Online Becomes a Problem." *BBC Newsbeat* (blog), April 2015. http://www.bbc.co.uk/newsbeat/article/32379961/cyber-stalking-when-looking-at-other-people-online-becomes-a-problem.

Thompson, Rachel. "What You Look Like May Determine Where You Are Seated in a Restaurant." *Mashable* (blog), January 2016. https://mashable.com/2016/01/05/restaurant-seating-appearance/.

Thonet, Thibaut, Guillaume Cabanac, Mohand Boughanem, and Karen Pinel-Sauvagnat. "Users Are Known by the Company They Keep: Topic Models for Viewpoint Discovery in Social Networks." In *Proceedings of the 2017 ACM on Conference on Information and Knowledge Management,* 87–96. CIKM '17. Association of Computing Machinery, 2017. https://doi.org/10.1145/3132847.3132897.

Topirceanu, Alexandru, Mihai Udrescu, and Radu Marculescu. "Weighted Betweenness Preferential Attachment: A New Mechanism Explaining Social Network Formation and Evolution." *Scientific Reports,* no. 8 (July 2018) 1–14. https://doi.org/10.1038/s41598-018-29224-w.

Trippi, Joe. "Technology Has Given Politics Back Its Soul." *MIT Technology Review* 116, no. 1 (2013) 34–36.

Tromholt, Morten. "The Facebook Experiment: Quitting Facebook Leads to Higher Levels of Well-Being." *Cyberpsychology, Behavior, and Social Networking* 19, no. 11 (November 2016) 661–66. https://doi.org/10.1089/cyber.2016.0259.

Tschabitscher, Heinz. "19 Fascinating Email Facts." *Lifewire* (blog), March 2020. https://www.lifewire.com/how-many-emails-are-sent-every-day-1171210.

Tunggal, Abi Tyas. "The 36 Biggest Data Breaches." *UpGuard* (blog), June 2020. https://www.upguard.com/blog/biggest-data-breaches.

Turel, Ofir. "Potential 'Dark Sides' of Leisure Technology Use in Youth." *Communications of the ACM* 62, no. 3 (February 2019) 24–27. https://doi.org/10.1145/3306615.

Turing, A. M. "Computing Machinery and Intelligence." *Mind, New Series* 59, no. 236 (1950) 433–60. www.jstor.org/stable/2251299.

———. "On Computable Numbers, with an Application to the Entscheidungs Problem." *Proceedings of the London Mathematical Society* 42, no. 2 (November 1936) 230–65.

Turkle, Sherry. *Alone Together*. New York: Basic, 2017.

———. *Life on the Screen: Identity in the Age of the Internet*. New York: Simon & Schuster, 2011.

Turner, Fred. *From Counterculture to Cyberculture: Stewart Brand, the Whole Earth Network, and the Rise of Digital Utopianism*. Kindle ed. Chicago: University of Chicago Press, 2006.

Turner, James. *Without God, Without Creed*. New Studies in American Intellectual and Cultural History. Baltimore: John Hopkins University Press, 1986.

Turow, Joseph, Michael Hennessy, and Nora Draper. "The Tradeoff Fallacy: How Marketers Are Misrepresenting American Consumers and Opening Them Up to Exploitation." Annenberg School for Communications, 2015. https://www.asc.upenn.edu/news-events/publications/tradeoff-fallacy-how-marketers-are-misrepresenting-american-consumers-and.

Turp, Michael-John. "Social Media, Interpersonal Relations and the Objective Attitude." *Ethics and Information Technology*, May 2020. https://doi.org/10.1007/s10676-020-09538-y.

Twenge, Jean M., Gabrielle N. Martin, and Brian H. Spitzberg. "Trends in U.S. Adolescents' Media Use, 1976–2016: The Rise of Digital Media, the Decline of TV, and the (near) Demise of Print." *Psychology of Popular Media Culture* 8, no. 4 (October 2019) 329–45. https://doi.org/10.1037/ppm0000203.

Underhill, Paco. *Why We Buy: The Science of Shopping*. New York: Simon & Schuster, 2000.

Vaidhyanathan, Siva. *Antisocial Media: How Facebook Disconnects Us and Undermines Democracy*. Oxford: Oxford University Press, 2018.

Valenzuela, Sebastián, Daniel Halpern, and James E. Katz. "Social Network Sites, Marriage Well-Being, and Divorce: Survey and State-Level Evidence from the United States." *Computers in Human Behavior* 36 (July 2014) 94–101. https://doi.org/10.1016/j.chb.2014.03.034.

Valjak, Ana. "Dark Patterns—Designs That Pull Evil Tricks on Our Brains." *The Capsized Eight* (blog), January 2019. https://infinum.co/the-capsized-eight/dark-patterns-designs-that-pull-evil-tricks-on-our-brains.

Vardi, Moshe Y. "The Winner-Takes-All Tech Corporation." *Communications of the ACM* 62, no. 11 (November 2018) 7. https://cacm.acm.org/magazines/2019/11/240377-the-winner-takes-all-tech-corporation/fulltext.

Veblen, Thorstein. "The Place of Science in Modern Civilization." *The American Journal of Sociology* 11, no. 5 (March 1906) 585–609.

———. *The Theory of Business Enterprise*. New York: C. Scribner's Sons, 1904.

Vermeeren, Arnold P. O. S., Effie Lai-Chong Law, Virpi Roto, Marianna Obrist, Jettie Hoonhout, and Kaisa Väänänen-Vainio-Mattila. "User Experience Evaluation Methods: Current State and Development Needs." In *Proceedings of the 6th Nordic Conference on Human-Computer Interaction: Extending Boundaries*, 521–30. NordiCHI '10. Reykjavik, Iceland: Association for Computing Machinery, 2010. https://doi.org/10.1145/1868914.1868973.

Victorino, R. C. "Embracing Asynchronous Communication in the Workplace." *Slab* (blog), April 2020. https://slab.com/blog/asynchronous-communication/.

Virus, Wuhan. "Despite Polling, Cell Phone Data Shows Americans Are Going Out Again." *The Federalist* (blog), May 2020. https://thefederalist.com/2020/05/06/despite-polling-cell-phone-data-shows-americans-are-going-out-again/.

Vondrackova, Petra, and David Smahel. "Internet Addiction." In *The Wiley Handbook of Psychology, Technology, and Society*, edited by Larry D. Rosen, Nancy A. Chaveer, and L. Mark Carrier, 469–85. Malden, MA: Wiley-Blackwell, 2015.

Wadhwa, Vivek. "Workplace Technology Is as Addictive as a Casino's Slot Machine—and Makes Us Less Productive." *MarketWatch* (blog), July 2018. https://www.marketwatch.com/story/workplace-technology-is-as-addictive-as-a-casinos-slot-machine-and-makes-us-less-productive-2018-07-30.

Wagner, Alan R., Jason Borenstein, and Ayanna Howard. "Overtrust in the Robotic Age." *Communications of the ACM* 61, no. 9 (August 2018) 22–24. https://doi.org/10.1145/3241365.

Waldman, Peter, Lizette Chapman, and Jordan Robertson. "Palantir Knows Everything About You." *Bloomberg* (blog), April 19, 2018. https://www.bloomberg.com/features/2018-palantir-peter-thiel/.

Waldron, Jeremy. "How Law Protects Dignity." *The Cambridge Law Journal* 71, no. 1 (2012) 200–222. https://www.jstor.org/stable/23253794.

Waters, Brent. *Christian Moral Theology in the Emerging Technoculture: From Posthuman Back to Human*. Burlington, VT: Ashgate, 2014.

———. *From Human to Posthuman: Christian Theology and Technology in a Postmodern World*. Burlington, VT: Ashgate, 2013.

Waters, Richard. "FT Interview with Google Co-Founder and CEO Larry Page." October 2014. https://www.ft.com/content/3173f19e-5fbc-11e4-8c27-00144feabdco.

Weber, Max. *The Protestant Ethic and Spirit of Capitalism*. Translated by Talcott Parsons. New York: Scribner's Sons, 1958.

Weckert, Simon. "Google Maps Hacks." http://www.simonweckert.com/googlemapshacks.html.

Wendel, Stephen. *Designing for Behavior Change: Applying Psychology and Behavioral Economics*. Kindle ed. Sebastopol, CA: O'Reilly Media, 2013.

Westin, Alan F., and Daniel J. Solove. *Privacy and Freedom*. Kindle ed. New York: Ig, 2018.

White, Russ. "Death of Transit: A Need to Prevent Fragmentation." *CircleID* (blog), November 2016. http://www.circleid.com/posts/20161107_death_of_transit_need_to_prevent_fragmentation/.

White, Russ, Tom Ammon, and Geoff Huston. "The Hedge 6: Geoff Huston on DoH." https://rule11.tech/the-hedge-episode-6-geoff-huston/.

Wickel, Taylor M. "Narcissism and Social Networking Sites: The Act of Taking Selfies." *The Elon Journal of Undergraduate Research in Communications* 6, no. 1 (Spring 2015) 5–12. http://www.inquiriesjournal.com/a?id=1138.

Wicker, Stephen B., and Dipayan Ghosh. "Reading in the Panopticon: Your Kindle May Be Spying on You, But You Can't Be Sure." *Communications of the ACM* 63, no. 5 (May 2020) 68–73. https://doi.org/10.1145/3376899.

Wiener, Norbert. *The Human Use of Human Beings: Cybernetics and Society*. Boston: Da Capo, 1988.

Wihbey, John, Kenneth Joseph, and David Lazer. "The Social Silos of Journalism? Twitter, News Media and Partisan Segregation." *New Media & Society* 21, no. 4 (2019) 815–35. https://doi.org/10.1177/1461444818807133.

Williams, Elisha. *The Essential Rights and Liberties of Protestants: A Seasonable Plea for the Liberty of Conscience, and the Right of Private Judgment, in Matters of Religion.* Boston: Kneeland & Green, 1744.

Williams, James. *Stand Out of Our Light.* Cambridge: Cambridge University Press, 2018.

Wilson, Timothy, David Reinhard, Erin Westgate, Daniel Gilbert, Nicole Ellerbeck, Cheryl Hahn, Casey Brown, and Adi Shaked. "Just Think: The Challenges of the Disengaged Mind." *Science* 345 (July 2014) 75–77. https://doi.org/10.1126/science.1250830.

Winner, Langdon. *The Whale and the Reactor: A Search for Limits in an Age of High Technology.* 1st ed. Chicago: University of Chicago Press, 1989.

Winther, Daniel Kardefelt. "How Does the Time Children Spend Using Digital Technology Impact Their Mental Well-Being, Social Relationships and Physical Activity? An Evidence-Focused Literature Review." UNICEF, December 2017. https://www.unicef-irc.org/publications/925-how-does-the-time-children-spend-using-digital-technology-impact-their-mental-well.html.

Wolfe, Tom. "Sorry, But Your Soul Just Died." *Orthodoxy Today* (blog), 2003. http://www.orthodoxytoday.org/articles/Wolfe-Sorry-But-Your-Soul-Just-Died.php.

Wood, Wendy, and David T. Neal. "The Habitual Consumer." *Journal of Consumer Psychology* 19, no. 4 (2009) 579–92. https://www.jstor.org/stable/45105782.

Wu, Tim. *The Attention Merchants: The Epic Scramble to Get Inside Our Heads.* New York: Vintage, 2016.

Xu, Fengli, Zhen Tu, Yong Li, Pengyu Zhang, Xiaoming Fu, and Depeng Jin. "Trajectory Recovery from Ash." *Proceedings of the 26th International Conference on World Wide Web,* 2017, 1241–50. https://doi.org/10.1145/3038912.3052620.

Xu, Luyi. "Exploiting Psychology and Social Behavior for Game Stickiness." *Communications of the ACM* 61, no. 11 (October 2018) 52–53. https://doi.org/10.1145/3239544.

Zacks Investment Research. "Facebook, Inc. (FB) Market Cap." February 2020. https://www.zacks.com/stock/chart/FB/fundamental/market-cap.

Zarsky, Tal. "Transparent Predictions." *University of Illinois Law Review* 2013.4 (2013) 1503–70. https://papers.ssrn.com/sol3/papers.cfm?abstract_id=2324240.

Zhao, Shanyang, Sherri Grasmuck, and Jason Martin. "Identity Construction on Facebook." *Computers in Human Behavior* 24, no. 5 (September 2008) 1816–36. https://doi.org/10.1016/j.chb.2008.02.012.

Zhihui, Cao. "Nowhere to Hide: Building Safe Cities with Technology Enablers and AI." *Huawei Publications* (blog), July 2016. https://www.huawei.com/us/about-huawei/publications/winwin-magazine/AI/nowhere-to-hide.

Zuboff, Shoshana. *The Age of Surveillance Capitalism: The Fight for a Human Future at the New Frontier of Power.* New York: PublicAffairs, 2019.

Zuckerberg, Mark. "Bringing the World Closer Together." Facebook, June 2017. https://www.facebook.com/notes/mark-zuckerberg/bringing-the-world-closer-together/10154944663901634/.